Kenichi Shimizu

Bootstrapping Stationary ARMA-GARCH Models

VIEWEG+TEUBNER RESEARCH

Kenichi Shimizu

Bootstrapping Stationary ARMA-GARCH Models

With a foreword by Prof. Dr. Jens-Peter Kreiß

VIEWEG+TEUBNER RESEARCH

Bibliographic information published by the Deutsche Nationalbibliothek
The Deutsche Nationalbibliothek lists this publication in the Deutsche Nationalbibliografie; detailed bibliographic data are available in the Internet at http://dnb.d-nb.de.

Dissertation Technische Universität Braunschweig, 2009

1st Edition 2010

Editorial Office: Dorothee Koch | Anita Wilke

Vieweg+Teubner is part of the specialist publishing group Springer Science+Business Media.
www.viewegteubner.de

Cover design: KünkelLopka Medienentwicklung, Heidelberg
Printing company: STRAUSS GMBH, Mörlenbach
Printed on acid-free paper
Printed in Germany

ISBN 978-3-8348-0992-6

This book is dedicated to my grandmother Sumi Yamazaki,

who always teaches me effort, tolerance and optimism.

Geleitwort

Im Jahre 1979 hat Bradley Efron mit seiner Arbeit *Bootstrap Methods: Another Look at the Jackknife* das Tor zu einem in den vergangenen 30 Jahren intensiv bearbeiteten Forschungsgebiet aufgestoßen. Die simulationsbasierte Methode des Bootstraps hat sich in den verschiedensten Bereichen als ein außerordentlich effizientes Werkzeug zur Approximation der stochastischen Fluktuation eines Schätzers um die zu schätzende Größe erwiesen. Präzise Kenntnis dieser stochastischen Fluktuation ist zum Beispiel notwendig, um Konfidenzbereiche für Schätzer anzugeben, die die unbekannte interessierende Größe mit einer vorgegebenen Wahrscheinlichkeit von, sagen wir, 95 oder 99% enthalten. In vielen Fällen und bei korrekter Anwendung ist das Bootstrapverfahren dabei der konkurrierenden und auf der Approximation durch eine Normalverteilung basierenden Methode überlegen. Die Anzahl der Publikationen im Bereich des Bootstraps ist seit 1979 in einem atemberaubenden Tempo angestiegen. Die wesentliche und im Grunde einfache Idee des Bootstraps ist die Erzeugung vieler (Pseudo-) Datensätze, die von ihrer wesentlichen stochastischen Struktur dem Ausgangsdatensatz möglichst ähnlich sind.

Die aktuellen Forschungsinteressen im Umfeld des Bootstraps bewegen sich zu einem großen Teil im Bereich der stochastischen Prozesse. Hier stellt sich die zusätzliche Herausforderung, bei der Erzeugung die Abhängigkeitsstruktur der Ausgangsdaten adäquat zu imitieren. Dabei ist eine präzise Analyse der zugrunde liegenden Situation notwendig, um beurteilen zu können, welche Abhängigkeitsaspekte für das Verhalten der Schätzer wesentlich sind und welche nicht, um ausreichend komplexe, aber eben auch möglichst einfache Resamplingvorschläge für die Erzeugung der Bootstrapdaten entwickeln zu können.

Herr Shimizu hat sich in seiner Dissertation die Aufgabe gestellt, für die Klasse der autoregressiven (AR) und autoregressiven moving average (ARMA) Zeitreihen mit bedingt heteroskedastischen Innovationen adäquate Bootstrapverfahren zu untersuchen. Bedingte Heteroskedastizität im Bereich von Zeitreihen ist ein überaus aktuelles Forschungsthema, da insbesondere Finanzdatensätze als ein wesentliches Merkmal diese Eigenschaft tragen. Zur Modellierung der bedingten Heteroskedastizität unterstellt der Autor die Modellklasse der ARCH- und GARCH-Modelle, die von Robert Engle initiiert und im Jahr 2003 mit dem renommierten Preis für Wirtschaftswissenschaften der schwedischen Reichsbank in Gedenken an Alfred Nobel ausgezeichnet wurde.

Herr Shimizu stellt zu Beginn seiner Arbeit mit einem negativen Beispiel sehr schön dar, welchen Trugschlüssen man bei der Anwendung des Bootstraps unterliegen kann. Danach untersucht er konsequent zwei verschiedene Modellklassen

(AR-Modelle mit ARCH-Innovationen und ARMA-Modelle mit GARCH-Innovationen) im Hinblick auf die Anwendbarkeit verschiedener Bootstrapansätze. Er beweist einerseits asymptotisch, d.h. für gegen unendlich wachsenden Stichprobenumfang, geltende Resultate. Andererseits dokumentiert er in kleinen Simulationsstudien die Brauchbarkeit der vorgeschlagenen Methoden auch für mittlere Stichprobenumfänge.

Im letzten Kapitel seiner Arbeit geht der Autor dann noch einen deutlichen Schritt weiter. Hier wird die parametrische Welt verlassen und ein semiparametrisches Modell betrachtet. Es geht darum, einen unbekannten funktionalen und möglicherweise nichtlinearen Zusammenhang bezüglich der nichtbeobachtbaren Innovationen zu schätzen und dann in einem weiteren Schritt die Verteilung des Kurvenschätzers mit einem Bootstrap-Ansatz zu imitieren.

Die vorliegende Arbeit, die am Institut für Mathematische Stochastik der Technischen Universität Braunschweig angefertigt wurde, erweitert die Möglichkeiten der Anwendung des Bootstrapverfahrens auf eine Klasse von autoregressiven Moving-Average-Zeitreihen mit heteroskedastischen Innovationen. Dabei werden die Verfahren nicht nur hinsichtlich ihrer asymptotischen Qualität, sondern auch in Simulationen für mittlere Stichprobenumfänge erprobt. Hierdurch werden nicht nur dem ambitionierten Anwender wertvolle Hinweise für einen adäquaten Ansatz des Bootstraps in sehr komplexen Situationen mit abhängigen Daten gegeben. Ich bin davon überzeugt, dass der Bereich des Bootstraps und allgemeiner des Resamplings für abhängige Daten noch für längere Zeit ein Feld intensiver akademischer Forschung und gleichzeitig fortgeschrittener praktischer Anwendung in zahlreichen Bereichen bleiben wird.

Prof. Dr. Jens-Peter Kreiß
Technische Universität Braunschweig

Acknowledgements

I am indebted to my PhD supervisor Professor Dr. Jens-Peter Kreiß for proposing this interesting research topic and for many valuable comments. Without his constructive criticism this book would not exist. I also thank Dr. Frank Palkowski, Ingmar Windmüller and Galina Smorgounova for helpful discussions on the topic. I would like to acknowledge the generous support of the Ministry of Finance Japan, which enabled me to continue working in this research area.

Last but not least, heartily thanks to Katrin Dohlus for sharing fun and pain in the last part of our dissertations and publications, and for her continuous enthusiastic encouragement. Very special thanks to my grandmother, parents and brother for their constant support and encouragement throughout this work.

Kenichi Shimizu

Contents

List o igures

1 Introduction

1.1 Financial Time Series and the GARCH Model

In financial time series analysis it is difficult to handle the observed discrete time asset price data P_t directly, because P_t are often nonstationary and highly correlated (Figure 1.1 (a)). Thus it is usual to analyse the *rate of return* on the financial instrument, which is generally stationary and uncorrelated.

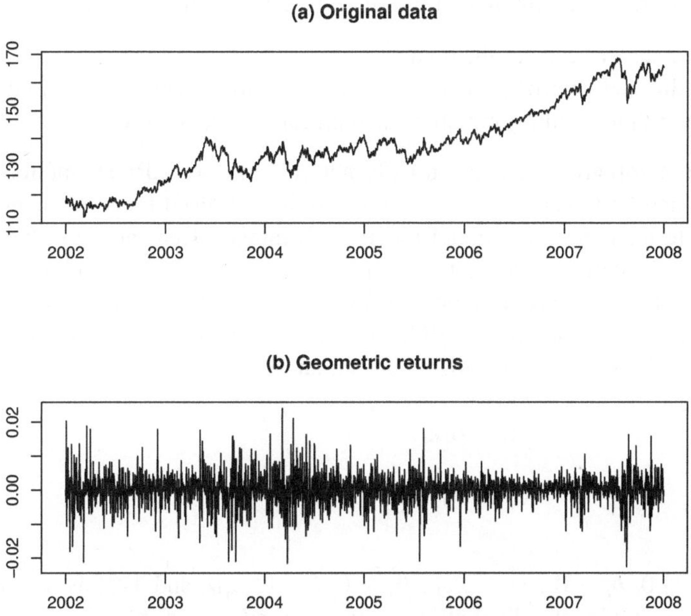

Figure 1.1: Euro to Yen exchange rate (1st January 2002 - 31st December 2007)

Here it is common to use *geometric returns*

$$X_t = \log\left(\frac{P_t}{P_{t-1}}\right) = \log P_t - \log P_{t-1}$$

(Figure 1.1 (b)) because of the property that compounded geometric returns are given as sums of geometric returns and that geometric returns are approximately equal to *arithmetic returns*

$$r_t = \frac{P_t - P_{t-1}}{P_{t-1}} = \frac{P_t}{P_{t-1}} - 1.$$

Emprical studies on financial time series[1] show that it is characterised by

- serial dependence in the data,
- conditional heteroscedasticity, i.e. the volatility changes over time,
- heavy-tailed and asymmetric unconditional distribution.

It was the autoregressive conditionally heteroscedastic (ARCH) model of Engle (1982) which first succeeded in capturing those "stylised facts of financial data" in a simple manner. The model has the advantage of easiness in obtaining the theoretical properties of interest, such as stationarity conditions and moments, and simplicity of numerical estimation. A few years later it was generalised by Bollerslev (1986) to the generalised ARCH (GARCH) model. A stochastic process $\{X_t\}$ is called GARCH(p, q) if it satisfies the equation

$$X_t = \sigma_t Z_t,$$

$$\sigma_t^2 = b_0 + \sum_{i=1}^{p} b_i X_{t-i}^2 + \sum_{j=1}^{q} \beta_j \sigma_{t-j}^2,$$

where $b_0 > 0$, $b_i \geq 0, i = 1,...,p$, $\beta_j \geq 0, j = 1,...,q$, and $\{Z_t\}$ is a sequence of independent and identically distributed (i.i.d.) random variables such that $\mathscr{E}(Z_t) = 0$ and $\mathscr{E}(Z_t^2) = 1$. The process is called ARCH(p) if $q = 0$. After this important discovery a variety of alternative conditionally heteroscedastic models have been proposed, however, the GARCH model is still the benchmark model of financial time series analysis because of its simplicity.[2]

[1] See, e.g., Mandelbrot (1963), Fama (1965) and Straumann (2005).

[2] See Bollerslev, Chou and Kroner (1992) for the application of ARCH models to financial time series data.

1.2 The Limit of the Classical Statistical Analysis

In the classical statistical analysis we obtain asymptotic properties of the GARCH estimators by the central limit theorem for martingale difference sequence.[3] It is important to note that the theorem yields the *asymptotic* properties of the estimators. In practice, however, the number of observations is limited and thus the error in the normal approximation not negligible. In general, the smaller the number of data is, the larger is the approximation error. While in simpler models one obtains a good approximation with small number of observations, more complex models such as the ARMA-GARCH require large amounts of data for a satisfactory approximation. In short, by the classical statistical analysis it is possible to obtain an estimator for the GARCH model, but often difficult to analyse how accurate the estimator is.

Example 1.1 (ARMA(1,1)-GARCH(1,1) Model)
Let us consider the ARMA(1,1)-GARCH(1,1) model:

$$\begin{aligned} X_t &= a_0 + a_1 X_{t-1} + \alpha_1 \varepsilon_{t-1} + \varepsilon_t, \\ \varepsilon_t &= \sqrt{h_t}\eta_t, \\ h_t &= b_0 + b_1 \varepsilon_{t-1}^2 + \beta_1 h_{t-1}, \qquad t = 1,...,2,000. \end{aligned}$$

We simulate now the error in the normal approximation. The simulation procedure is as follows.

(step 1) Simulate the bias $\eta_t \overset{iid}{\sim} \sqrt{\frac{3}{5}}\, t_5$ for $t = 1,...,2,000$.

(step 2) Let $X_0 = \mathscr{E}(X_t) = \frac{a_0}{1-a_1}$, $h_0 = \varepsilon_0^2 = \mathscr{E}(\varepsilon_t^2) = \frac{b_0}{1-b_1-\beta}$, $a_0 = 0.141$, $a_1 = 0.433$, $\alpha_1 = -0.162$, $b_0 = 0.007$, $b_1 = 0.135$ and $\beta_1 = 0.829$ and calculate

$$X_t = a_0 + a_1 X_{t-1} + \alpha_1 \varepsilon_{t-1} + \left(b_0 + b_1 \varepsilon_{t-1}^2 + \beta_1 h_{t-1}\right)^{1/2} \eta_t$$

for $t = 1,...,2,000$.

(step 3) From the data set $\{X_1, X_2, ..., X_{2,000}\}$ obtain the quasi-maximum likelihood (QML) estimators

$$\widehat{\boldsymbol{\theta}} = (\widehat{a_0}\ \widehat{a_1}\ \widehat{\alpha_1}\ \widehat{b_0}\ \widehat{b_1}\ \widehat{\beta_1})'.$$

(step 4) Repeat (step 1)-(step 3) 1,000 times.

---

[3] See chapter 4 for the further discussion.

(step 5) Compute the estimated variance $\widehat{\Sigma}^{-1}\,\widehat{\Omega}\,\widehat{\Sigma}^{-1}$,[4] where

$$\widehat{\Sigma} = \frac{1}{T}\sum_{t=1}^{T}\frac{1}{2h_t^2(\widehat{\boldsymbol{\theta}})}\left(\frac{\partial h_t(\widehat{\boldsymbol{\theta}})}{\partial b_0}\right)^2 \quad \textit{and}$$

$$\widehat{\Omega} = \frac{1}{2}\left(\frac{1}{T}\sum_{t=1}^{T}\frac{\varepsilon_t^4}{h_t^2}(\widehat{\boldsymbol{\theta}}) - 1\right)\frac{1}{T}\sum_{t=1}^{T}\frac{1}{2h_t^2(\widehat{\boldsymbol{\theta}})}\left(\frac{\partial h_t(\widehat{\boldsymbol{\theta}})}{\partial b_0}\right)^2.$$

(step 6) Compare the simulated density of the QML estimator ($\sqrt{T}(\widehat{b_0} - b_0)$, thin line) with its normal approximation ($\mathcal{N}\left(0,\ \widehat{\Sigma}^{-1}\,\widehat{\Omega}\,\widehat{\Sigma}^{-1}\right)$, bold line).

As we see in Figure 1.2, the error is fairly large because the number of data is small.

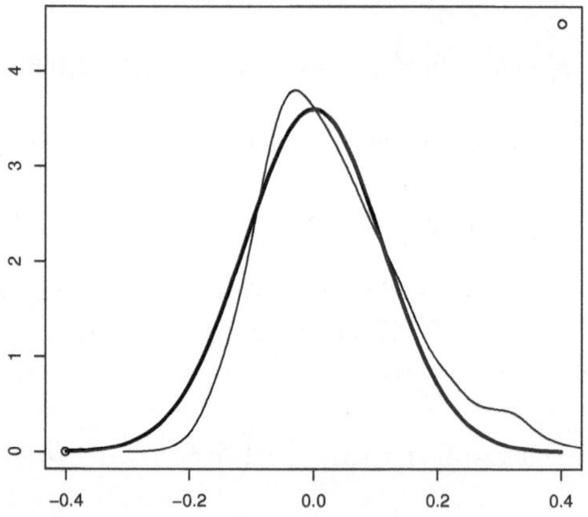

Figure 1.2: $\sqrt{T}(\widehat{b_0} - b_0)$ and $\mathcal{N}\left(0,\ \widehat{\Sigma}^{-1}\,\widehat{\Omega}\,\widehat{\Sigma}^{-1}\right)$

[4] See Theorem 4.1 and 4.2 for the asymptotic properties of the QML estimators for the ARMA(p, q)-GARCH(r, s) model.

1.3 An Alternative Approach: the Bootstrap Techniques

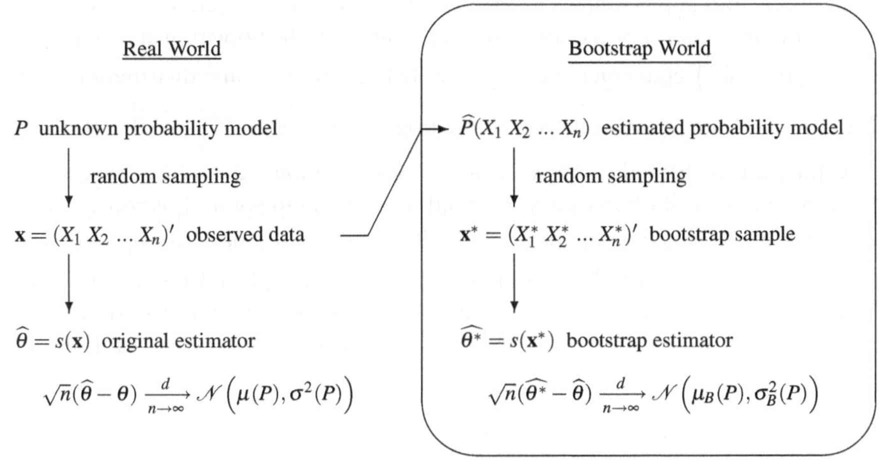

Figure 1.3: The basic idea of the bootstrap techniques

There is another approach to analyse the accuracy of the estimators. Efron (1979) proposed the bootstrap technique, in which one estimates the accuracy of an estimator by computer-based replications of the estimator. Figure 1.3 is a schematic diagram of the bootstrap technique.[5] In the real world an unknown probability model P generates the observed data $\mathbf{x}$ by random sampling. We calculate the estimator $\widehat{\theta}$ of interest from $\mathbf{x}$. The statistical properties of $\widehat{\theta}$ is obtained by the central limit theorem

$$\sqrt{n}(\widehat{\theta} - \theta) \xrightarrow[n\to\infty]{d} \mathcal{N}\left(\mu(P),\ \sigma^2(P)\right),$$

where we estimate unknown $\mu(P)$ and $\sigma^2(P)$ by their consistent estimators $\widehat{\mu}(\mathbf{x})$ and $\widehat{\sigma^2}(\mathbf{x})$. On the contrary, in the bootstrap world we obtain the estimated probability model $\widehat{P}$ from $\mathbf{x}$. Then $\widehat{P}$ generates the bootstrap samples $\mathbf{x}^*$ by random sampling, from which we calculate the bootstrap estimator $\widehat{\theta^*}$. Here the mapping from $\mathbf{x}^* \to \widehat{\theta^*}$, $s(\mathbf{x}^*)$, is the same as the mapping from $\mathbf{x} \to \widehat{\theta}$, $s(\mathbf{x})$. Note

[5] The diagram is a modification of Figure 8.3 in Efron and Tibshirani (1993, p. 91).

that in the real world we have only one data set $\mathbf{x}$ and thus can obtain one estimator $\widehat{\theta}$, while in the bootstrap world it is possible to calculate $\mathbf{x}^*$ and $\widehat{\theta^*}$ as many times as we can afford. That is, through a large number of bootstrap replications we can approximate the estimation error of the original estimator $\sqrt{n}(\widehat{\theta}-\theta)$ by the bootstrap approximation $\sqrt{n}(\widehat{\theta^*}-\widehat{\theta})$. A bootstrap technique is called to *work* (or to be *weakly consistent*) if the distribution of the bootstrap approximation $\mathscr{L}\left(\sqrt{n}(\widehat{\theta^*}-\widehat{\theta})\right)$ converges weakly in probability to the same distribution as the distribution of the estimation error of the original estimator $\mathscr{L}\left(\sqrt{n}(\widehat{\theta}-\theta)\right)$.

Although it is difficult to prove it theoretically, empirical studies[6] have shown that if the number of observations is small, the bootstrap approximation $\sqrt{n}(\widehat{\theta^*}-\widehat{\theta})$ yields a better approximation to $\sqrt{n}(\widehat{\theta}-\theta)$ than the normal approximation $\mathscr{N}(\widehat{\mu}(\mathbf{x}), \widehat{\sigma^2}(\mathbf{x}))$. Figure 1.4 also indicates that in Example 1.1 the residual bootstrap approximation (see section 4.2 and 4.4 for the simulation procedures) provides a slightly better approximation than the normal approximation (cf. Figure 1.2).

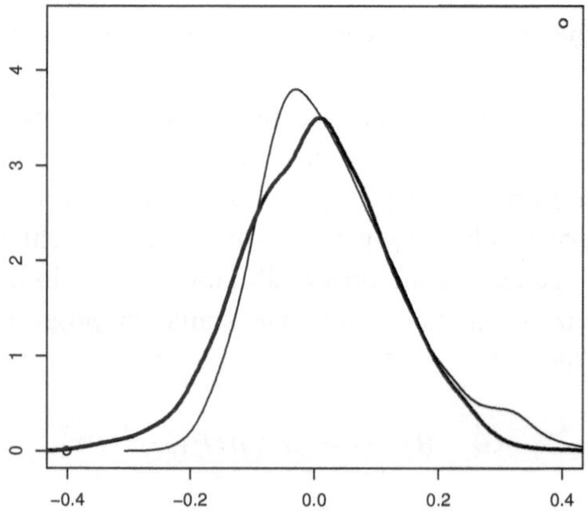

Figure 1.4: $\sqrt{T}(\widehat{b_0}-b_0)$ and $\sqrt{T}(\widehat{b_0^*}-\widehat{b_0})$

[6] See, e.g., Horowitz (2003), Pascual, Romo and Ruiz (2005), Reeves (2005), Robio (1999), and Thombs and Schucany (1990).

Here it is important to note that a bootstrap technique does not always work. Whether a bootstrap works, depends on the type of bootstrap procedures, probability models and estimation methods. Especially in presence of conditional heteroscedasticity, which is one of the main characteristics of financial time series, naïve application of bootstrap could result in a false conclusion, as we will see in the following chapter. From this viewpoint this work aims to research the asymptotic properties of the two standard bootstrap techniques - the residual and the wild bootstrap - for the stationary ARMA-GARCH models.

1.4 Structure of the Book

This work is motivated by the following three research results: First, Kreiß (1997) investigated the AR(p) model with heteroscedastic errors and proved weak consistency of the residual and the wild bootstrap techniques. Second, Maercker (1997) showed weak consistency of the wild bootstrap for the GARCH(1,1) model with the quasi-maximum likelihood estimation. Third, Franke, Kreiß and Mammen (2002) examined the NARCH(1) model and proved weak consistency of the residual and the wild bootstrap techniques. In this work we sketch the results and try to extend them to the general bootstrap theory for heteroscedastic models.

This book is organised as follows. Following a brief introduction to the heteroscedastic time series models, chapter 2 explains potential problems of bootstrap methods for heteroscedastic models and points out the risk of a false application with the example of the VaR theory. The third chapter treats the parameter estimation for the parametric AR(p)-ARCH(q) models with the ordinary least squares estimation. For the AR(p)-ARCH(q) models the residual and the wild bootstrap techniques are adopted and the consistency of the estimators investigated. For both methods a suitable procedure and the consistency criterium are determined. The theoretical analysis is then confirmed by simulations. In the following two chapters the results of chapter 3 are extended to two further models, which are the parametric ARMA(p, q)-GARCH(r, s) models with the quasi-maximum likelihood estimation on the one hand and the semiparametric AR(p)-ARCH(1) models with the Nadaraya-Watson estimation, in which heteroscedasticity parameters are nonparametric, on the other. For both models a suitable procedure and the consistency criterium are determined. Then the analysis is confirmed by simulations.

2 Bootstrap Does not Always Work

As we have seen in the previous chapter, bootstrap is without doubt a promising technique, which supplements some weak points of classical statistical methods. In empirical studies, however, the limit of bootstrap tends to be underestimated, and the technique is sometimes regarded as a utility tool applicable to all models. Let us see a typical misunderstanding of bootstrap in econometric literature.

> The bootstrap is a simple and useful method for assessing uncertainty in estimation procedures. Its distinctive feature is that it replaces mathematical or statistical analysis with simulation-based resampling from a given data set. It therefore provides a means of assessing the accuracy of parameter estimators without having to resort to strong parametric assumptions or closed-form confidence-interval formulas. (...) The bootstrap is also easy to use because it does not require the user to engage in any difficult mathematical or statistical analysis. In any case, such traditional methods only work in a limited number of cases, whereas the bootstrap can be applied more or less universally. So the bootstrap is easier to use, more powerful and (as a rule) more reliable than traditional means of estimating confidence intervals for parameters of interest.[1]

It is important to note, however, that bootstrap does not always work. As mentioned in the previous chapter, whether a bootstrap method works, depends not only on the type of bootstrap technique, but also on the chosen model and the selected estimation method. In fact, we always need to check whether the bootstrap technique combined with the statistical model and the estimation method works or not. From this viewpoint it is naïve to believe that bootstrap does not require mathematical or statistical analysis.

When does bootstrap work? When does bootstrap fail to work? Classical examples are explained in Mammen (1992). In this paper we concentrate on the case where we estimate heteroscedasticity, which is indispensable for financial risk management. As we will see, when estimating heteroscedasticity, the third

[1] Dowd (2005, p. 105).

and fourth moment of the bootstrap residuals play a substantial role, which requires more careful attention when applying bootstrap.

In the first section we see possible mistakes in the application of bootstrap when estimating heteroscedasticity. In the second section we observe how a false application of bootstrap could lead to a grave mistake in practice.

2.1 Estimation of Heteroscedasticity and Bootstrap

Assume that $\{X_t,\ t = 2,...,T\}$ are generated by the following AR(1)-ARCH(1) model:

$$\begin{aligned} X_t &= aX_{t-1} + \varepsilon_t, \\ \varepsilon_t &= \sqrt{h_t}\eta_t, \\ h_t &= b_0 + b_1\varepsilon_{t-1}^2, \end{aligned} \tag{2.1}$$

where $b_0 > 0$, $b_1 \geq 0$ and $\{\eta_t\}$ is a sequence of i.i.d. random variables such that $\mathscr{E}(\eta_t) = 0$, $\mathscr{E}(\eta_t^2) = 1$, following a symmetric distribution, i.e. $\mathscr{E}(\eta_t^3) = 0$, and $\mathscr{E}(\eta_t^4) =: \kappa < \infty$.[2] For simplicity, we adopt the ordinary least squares (OLS) method for parameter estimation throughout this section. Note that the OLS estimation consists of two steps, in which the ARCH part is estimated based on the residuals

$$\widehat{\varepsilon}_t = X_t - \widehat{a}X_{t-1}, \quad t = 2,\ ...,\ T.$$

Now we apply the residual and the wild bootstrap to the AR(1)-ARCH(1) model and see the cases where bootstrap fails to work.

2.1.1 Simple Residual Bootstrap Does not Work

The simplest residual bootstrap method is carried out as follows:[3]

(step 1) Obtain the OLS estimator $\widehat{a}$ and calculate the residuals

$$\widehat{\varepsilon}_t = X_t - \widehat{a}X_{t-1}, \quad t = 2,...,T.$$

(step 2) Compute the standardised residuals

$$\widetilde{\varepsilon}_t = \widehat{\varepsilon}_t - \widehat{\mu}$$

[2] For further assumptions see §3.1.1.

[3] See, e.g., Efron and Tibshirani (1993, p. 111), Shao and Tu (1996, p. 289) and Kreiß (1997).

for $t = 2, ..., T$, where

$$\widehat{\mu} = \frac{1}{T-1} \sum_{t=2}^{T} \widehat{\varepsilon}_t .$$

(step 3) Obtain the empirical distribution function $\mathscr{F}_T(x)$ based on $\widetilde{\varepsilon}_t$ defined by

$$\mathscr{F}_T(x) := \frac{1}{T-1} \sum_{t=2}^{T} \mathbf{1}\left(\widetilde{\varepsilon}_t \leq x\right).$$

(step 4) Generate the bootstrap process X_t^* by computing

$$X_t^* = \widehat{a} X_{t-1} + \varepsilon_t^*,$$
$$\varepsilon_t^* \overset{iid}{\sim} \mathscr{F}_T(x), \qquad t = 2, ..., T.$$

(step 5) Calculate the bootstrap estimator

$$\widehat{a^*} = (\mathbf{x}'\mathbf{x})^{-1}\mathbf{x}'\mathbf{x}^*,$$
$$\widetilde{\mathbf{b}^*} = (\widehat{\mathbf{E}}'\widehat{\mathbf{E}})^{-1}\widehat{\mathbf{E}}'\mathbf{e}^*,$$

where $\mathbf{x} = (X_1\ X_2\ ...\ X_{T-1})'$, $\mathbf{x}^* = (X_2^*\ X_3^*\ ...\ X_T^*)'$, $\widetilde{\mathbf{b}^*} = (\widetilde{b_0^*}\ \widetilde{b_1^*})'$, $\mathbf{e}^* = (\varepsilon_2^{*2}\ \varepsilon_3^{*2}\ ...\ \varepsilon_T^{*2})'$ and

$$\widehat{\mathbf{E}} = \begin{pmatrix} 1 & \widehat{\varepsilon_1}^2 \\ 1 & \widehat{\varepsilon_2}^2 \\ \vdots & \vdots \\ 1 & \widehat{\varepsilon_{T-1}}^2 \end{pmatrix}.$$

In this case we obtain

$$\begin{aligned} \mathscr{E}_*(\varepsilon_t^*) &= \sum_{t=2}^{T} (\widehat{\varepsilon}_t - \widehat{\mu}) \frac{1}{T-1} \\ &= \underbrace{\frac{1}{T-1} \sum_{t=2}^{T} \widehat{\varepsilon}_t}_{=\widehat{\mu}} - \widehat{\mu} \\ &= 0 \end{aligned}$$

and

$$\mathscr{E}_*(\varepsilon_t^{*2}) = \sum_{t=2}^{T}(\widehat{\varepsilon}_t - \widehat{\mu})^2 \frac{1}{T-1}$$
$$= \frac{1}{T-1}\sum_{t=2}^{T}\widehat{\varepsilon}_t^{\,2} - 2\widehat{\mu}\underbrace{\frac{1}{T-1}\sum_{t=2}^{T}\widehat{\varepsilon}_t}_{=\widehat{\mu}} + \widehat{\mu}^2$$
$$= \frac{1}{T-1}\sum_{t=2}^{T}\widehat{\varepsilon}_t^{\,2} - \widehat{\mu}^2$$
$$= \mathscr{E}(\varepsilon_t^2) - \Big(\underbrace{\mathscr{E}(\varepsilon_t)}_{=0}\Big)^2 + o_p(1),$$

where $\mathscr{E}_*$ denotes the conditional expectation given the observations $X_1, ..., X_T$. These properties yield weak consistency of the bootstrap estimators for the AR part (cf. the proof of Theorem 3.3). For the ARCH part, however, we do not obtain the necessary condition for weak consistency (cf. the proof of Theorem 3.4), namely

$$\mathscr{E}_*\Big(\varepsilon_t^{*2} - (\widetilde{b_0} + \widetilde{b_1}\widehat{\varepsilon_{t-1}}^2)\Big) \neq 0.$$

In short, this simplest technique works for the AR part of the model (2.1), but not for the ARCH part. The problem of this method is that the bootstrap residuals ε_t^{*2} ignore the heteroscedasticity of ε_t^2. To overcome the problem it is necessary to take a more complicated approach so that ε_t^{*2} immitate the heteroscedasticity of ε_t^2 correctly, which we will propose in section 3.2.

2.1.2 Wild Bootstrap Usually Does not Work

Analogously to the residual bootstrap, the simplest wild bootstrap method is carried out as follows:[4]

(step 1) Obtain the OLS estimator $\widehat{a}$ and calculate the residuals

$$\widehat{\varepsilon}_t = X_t - \widehat{a}X_{t-1}, \qquad t = 2, ..., T.$$

(step 2) Generate the bootstrap process $X_t^\dagger$ by computing

$$X_t^\dagger = \widehat{a}X_{t-1} + \varepsilon_t^\dagger,$$
$$\varepsilon_t^\dagger = \widehat{\varepsilon}_t w_t^\dagger, \quad w_t^\dagger \overset{iid}{\sim} \mathscr{N}(0,1), \quad t = 2, ..., T.$$

[4] See, e.g., Liu (1988) and Kreiß (1997).

(step 3) Calculate the bootstrap estimator

$$\widehat{a}^{\dagger} = (\mathbf{x}'\mathbf{x})^{-1}\mathbf{x}'\mathbf{x}^{\dagger},$$
$$\widetilde{\mathbf{b}}^{\dagger} = (\widehat{\mathbf{E}}'\widehat{\mathbf{E}})^{-1}\widehat{\mathbf{E}}'\mathbf{e}^{\dagger},$$

where $\mathbf{x}^{\dagger} = (X_2^{\dagger}\ X_3^{\dagger}\ ...\ X_T^{\dagger})'$, $\widetilde{\mathbf{b}}^{\dagger} = (\widetilde{b_0^{\dagger}}\ \widetilde{b_1^{\dagger}})'$ and $\mathbf{e}^{\dagger} = (\varepsilon_2^{\dagger 2}\ \varepsilon_3^{\dagger 2}\ ...\ \varepsilon_T^{\dagger 2})'$.

In this case we obtain

$$\mathscr{E}_{\dagger}(\varepsilon_t^{\dagger}) = \widehat{\varepsilon}_t \mathscr{E}_{\dagger}(w_t^{\dagger}) = 0$$

and

$$\mathscr{E}_{\dagger}(\varepsilon_t^{\dagger 2}) = \widehat{\varepsilon}_t^{\,2} \mathscr{E}_{\dagger}(w_t^{\dagger 2}) = \mathscr{E}(\varepsilon_t^2) + o_p(1),$$

where $\mathscr{E}_{\dagger}$ denotes the conditional expectation given the observations $X_1, ..., X_T$. These properties yield weak consistency of the bootstrap estimators for the AR part (cf. the proof of Theorem 3.5). For the ARCH part, however, we do not obtain the necessary condition for weak consistency (cf. the proof of Theorem 3.6), namely

$$\mathscr{E}_{\dagger}\left(\varepsilon_t^{\dagger 2} - (\widetilde{b}_0 + \widetilde{b}_1 \widehat{\varepsilon_{t-1}}^2)\right) = \widehat{\varepsilon}_t^{\,2} - (\widetilde{b}_0 + \widetilde{b}_1 \widehat{\varepsilon_{t-1}}^2) \neq 0$$

and

$$\mathscr{E}_{\dagger}(\varepsilon_t^{\dagger 4}) = 3\mathscr{E}(\varepsilon_t^4) + o_p(1).$$

There are two problems with the simplest method. Firstly, the bootstrap residuals $\varepsilon_t^{\dagger 2}$ do not immitate the heteroscedasticity of ε_t^2 accurately, which is analogous to the simplest residual bootstrap. Secondly, $\varepsilon_t^{\dagger 2}$ have larger variance than ε_t^2. As we will see in section 3.3, it is possible to overcome the first problem by taking a more complicated approach. It is important to note, however, that the second problem still remains even if we take the more complicated approach (cf. Theorem 3.6 and 4.4). In conclusion: Wild bootstrap usually does not work when estimating heteroscedasticity (cf. Remark 3.6-3.7).

2.2 Result of a False Application: the VaR Model

In this section we demonstrate the danger of a false application of bootstrap. Firstly, based on Hartz, Mittnik and Paolella (2006) we sketch the theory of applying bootstrap to the Value-at-Risk (VaR) forecast model. Then we simulate the result of a false application of bootstrap to the VaR forecast model.

2.2.1 VaR Model

The VaR is one of the most prominent measures of financial market risk, which "summarizes the worst loss over a target horizon that will not be exceeded with a given level of confidence".[5] As a measure of unexpected loss at some confidence level, the VaR has been used by financial institutions to compute the necessary buffer capital. Since the Basel Committee adopted the internal-models-approach (1996), which allows banks to use their own risk measurement models to determine their capital charge, a large number of new models have been established. Among them the (G)ARCH model of Engle (1982) and Bollerslev (1986) is one of the simplest but still successful methods to obtain the VaR.

Assume that $\{X_t,\ t = 1,\dots,T\}$ are generated by the following ARMA(p, q)-GARCH(r, s) model:

$$\begin{aligned} X_t &= a_0 + \sum_{i=1}^{p} a_i X_{t-i} + \sum_{j=1}^{q} \alpha_j \varepsilon_{t-j} + \varepsilon_t, \\ \varepsilon_t &= \sqrt{h_t}\,\eta_t, \\ h_t &= b_0 + \sum_{j=1}^{s} b_j \varepsilon_{t-j}^2 + \sum_{i=1}^{r} \beta_i h_{t-i}, \end{aligned}$$

where $b_0 > 0$, $b_j \geq 0, j = 1,\dots,s$, $\beta_i \geq 0, i = 1,\dots,r$, and $\{\eta_t\}$ is a sequence of i.i.d. random variables such that $\mathscr{E}(\eta_t) = 0$, $\mathscr{E}(\eta_t^2) = 1$, following a symmetric distribution, i.e. $\mathscr{E}(\eta_t^3) = 0$, and $\mathscr{E}(\eta_t^4) =: \kappa < \infty$.[6] The k-step-ahead VaR forecast is

$$\widehat{\xi_\lambda}(k,T) = \Phi^{-1}(\lambda, \widehat{X_{T+k}}, \widehat{h_{T+k}}),$$

where $\Phi^{-1}(\lambda, \mu, \sigma^2)$ denotes the $\lambda \times 100\%$ quantile of the the normal distribution with mean μ and variance σ^2,

$$\begin{aligned} \widehat{X_{T+k}} &= \widehat{a_0} + \sum_{i=1}^{p} \widehat{a_i} X_{T+k-i} + \sum_{j=1}^{q} \widehat{\alpha_j} \widehat{\varepsilon_{T+k-j}} \quad \textit{and} \\ \widehat{h_{T+k}} &= \widehat{b_0} + \sum_{j=1}^{s} \widehat{b_j} \widehat{\varepsilon_{T+k-j}}^2 + \sum_{i=1}^{r} \widehat{\beta_i} \widehat{h_{T+k-i}} \end{aligned}$$

with $X_t = \widehat{X_t}$, $\widehat{\varepsilon_t} = 0$ and $\widehat{\varepsilon_t}^2 = \widehat{h_t}$ for $t > T$.

---

[5] Jorion (2007, p. 17).

[6] For further assumptions see §4.1.1.

The k-step-ahead VaR forecast $\widehat{\xi_\lambda}(k,T)$ is a point estimator. It is possible to construct confidence intervals of $\widehat{\xi_\lambda}(k,T)$ through computing the bootstrapped k-step-ahead VaR forecast

$$\widehat{\xi_\lambda^*}(k,T) = \Phi^{-1}(\lambda, X_{T+k}^*, h_{T+k}^*),$$

where

$$X_{T+k}^* = \widehat{a_0^*} + \sum_{i=1}^{p} \widehat{a_i^*} X_{T+k-i} + \sum_{j=1}^{q} \widehat{\alpha_j^*} \widehat{\varepsilon_{T+k-j}} \quad \textit{and}$$

$$h_{T+k}^* = \widehat{b_0^*} + \sum_{j=1}^{s} \widehat{b_j^*} \widehat{\varepsilon_{T+k-j}}^2 + \sum_{i=1}^{r} \widehat{\beta_i^*} \widehat{h_{T+k-i}},$$

numerous times with different bootstrapped estimators $\widehat{a_0^*}$, $\widehat{a_1^*}$, ..., $\widehat{a_p^*}$, $\widehat{\alpha_1^*}$, ..., $\widehat{\alpha_q^*}$, $\widehat{b_0^*}$, $\widehat{b_1^*}$, ..., $\widehat{b_s^*}$ and $\widehat{\beta_1^*}$, ..., $\widehat{\beta_r^*}$.

2.2.2 Simulations

Let us consider the following ARMA(1,1)-GARCH(1,1) model:

$$\begin{aligned} X_t &= a_0 + a_1 X_{t-1} + \alpha_1 \varepsilon_{t-1} + \varepsilon_t, \\ \varepsilon_t &= \sqrt{h_t}\eta_t, \\ h_t &= b_0 + b_1 \varepsilon_{t-1}^2 + \beta_1 h_{t-1}, \qquad t = 1, ..., 10{,}000. \end{aligned}$$

The simulation procedure is as follows. Note that the model and (step 1)-(step 3) are identical with the simulations in section 4.4.

(step 1) Simulate the bias $\eta_t \overset{iid}{\sim} \sqrt{\frac{3}{5}}\, t_5$ for $t = 1, ..., 10{,}000$.

(step 2) Let $X_0 = \mathscr{E}(X_t) = \frac{a_0}{1-a_1}$, $h_0 = \varepsilon_0^2 = \mathscr{E}(\varepsilon_t^2) = \frac{b_0}{1-b_1-\beta}$, $a_0 = 0.141$, $a_1 = 0.433$, $\alpha_1 = -0.162$, $b_0 = 0.007$, $b_1 = 0.135$ and $\beta_1 = 0.829$ and calculate

$$X_t = a_0 + a_1 X_{t-1} + \alpha_1 \varepsilon_{t-1} + \left(b_0 + b_1 \varepsilon_{t-1}^2 + \beta_1 h_{t-1}\right)^{1/2} \eta_t$$

for $t = 1, ..., 10{,}000$.

(step 3) From the data set $\{X_1, X_2, ..., X_{10{,}000}\}$ obtain the QML estimators $\widehat{a_0}$, $\widehat{a_1}$, $\widehat{\alpha_1}$, $\widehat{b_0}$, $\widehat{b_1}$ and $\widehat{\beta_1}$.

(step 4) Compute the one-step-ahead VaR forecast

$$\widehat{\xi_\lambda}(1,T) = \Phi^{-1}(\lambda, \widehat{X_{T+1}}, \widehat{h_{T+1}})$$

with $\lambda = 0.01$ and $T = 10{,}000$.

(step 5) Adopt the residual bootstrap of section 4.2 and the wild bootstrap of section 4.3 based on the data set and the estimators of (step 3). Then compute the bootstrapped one-step-ahead VaR forecast

$$\widehat{\xi_\lambda^*}(1,T) = \Phi^{-1}(\lambda, X_{T+1}^*, h_{T+1}^*) \quad \textit{and}$$
$$\widehat{\xi_\lambda^\dagger}(1,T) = \Phi^{-1}(\lambda, X_{T+1}^\dagger, h_{T+1}^\dagger),$$

where

$$X_{T+1}^* = \widehat{a_0^*} + \widehat{a_1^*} X_T + \widehat{\alpha_1^*}\widehat{\varepsilon_T},$$
$$h_{T+1}^* = \widehat{b_0^*} + \widehat{b_1^*}\widehat{\varepsilon_T}^2 + \widehat{\beta_1^*}\widehat{h_T},$$

and

$$X_{T+1}^\dagger = \widehat{a_0^\dagger} + \widehat{a_1^\dagger} X_T + \widehat{\alpha_1^\dagger}\widehat{\varepsilon_T},$$
$$h_{T+1}^\dagger = \widehat{b_0^\dagger} + \widehat{b_1^\dagger}\widehat{\varepsilon_T}^2 + \widehat{\beta_1^\dagger}\widehat{h_T},$$

with $\lambda = 0.01$ and $T = 10{,}000$.

(step 6) Repeat (step 5) 2,000 times and construct a 90% confidence interval for $\widehat{\xi_{0.01}^*}(1,\ 10{,}000)$ and $\widehat{\xi_{0.01}^\dagger}(1,\ 10{,}000)$.

Figure 2.1 displays the density functions of the bootstrapped one-step-ahead VaR forecast with the residual bootstrap ($\widehat{\xi_{0.01}^*}(1,\ 10{,}000)$, thin line) and the wild bootstrap ($\widehat{\xi_{0.01}^\dagger}(1,\ 10{,}000)$, bold line). The 90% confidence interval for $\widehat{\xi_{0.01}^*}(1,\ 10{,}000)$ is $[-0.5752,\ -0.5600]$ while that for $\widehat{\xi_{0.01}^\dagger}(1,\ 10{,}000)$ is $[-0.5737,\ -0.5614]$. As we will see in section 4.4, the wild bootstrap does not work for the ARMA(1, 1)-GARCH(1, 1) model, which results in the apparently narrower confidence interval for $\widehat{\xi_{0.01}^\dagger}(1,\ 10{,}000)$ than that for $\widehat{\xi_{0.01}^*}(1,\ 10{,}000)$ in this simulation. That is, the wild bootstrap technique underestimates the existing risk and could lead a financial institution to the false decision to take the too much risk that is not covered by her capital charge.

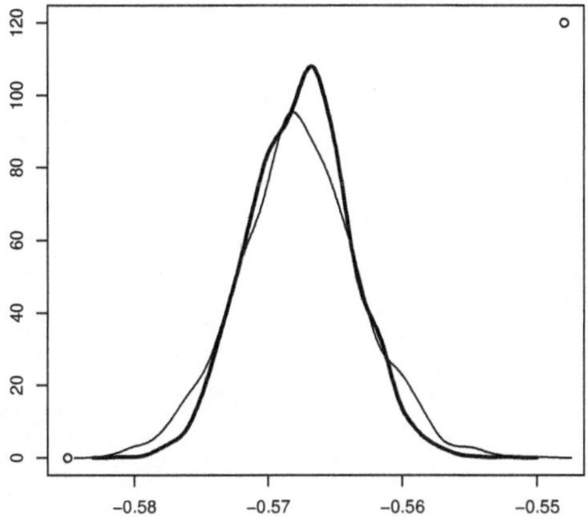

Figure 2.1: $\widehat{\xi^*_{0.01}}(1,\ 10{,}000)$ and $\widehat{\xi^\dagger_{0.01}}(1,\ 10{,}000)$

3 Parametric AR(p)-ARCH(q) Models

In this chapter we consider the parametric AR(p)-ARCH(q) model based on ARCH regression models of Engle (1982). For simplicity, we estimate the parameters by the ordinary least squares (OLS) method and adopt the two-step estimation for the ARCH part, in which the parameters of the ARCH part are estimated based on the residuals of the AR part. In the first section we sketch the estimation theory for the parametric AR(p)-ARCH(q) model with the OLS method and prove asymptotic normality of the OLS estimators. In the following two sections possible applications of the residual and the wild bootstrap are proposed and their weak consistency investigated. These theoretical results are confirmed by simulations in the last section.

3.1 Estimation Theory

Since Engle (1982) introduced the ARCH model, asymptotic theories for ARCH-type models have been presented by many researchers. Although most of them adopted the maximum likelihood method for parameter estimation, there are some studies on the OLS method. Weiss (1984, 1986) investigated the asymptotic theory for ARMA(p, q)-ARCH(s) models with the least squares estimation. Pantula (1988) examined the asymptotic properties of the generalised least squares estimators for the AR(p)-ARCH(1) model. In this section we consider the AR(p)-ARCH(q) model and sketch the asymptotic properties of the OLS estimators based on these past research results.

3.1.1 Model and Assumptions

Assume that $\{X_t,\ t=p+1,\ ...,\ T\}$ are generated by the p-th order autoregressive model with the ARCH(q) errors:

$$\begin{aligned} X_t &= a_0 + \sum_{i=1}^{p} a_i X_{t-i} + \varepsilon_t, \\ \varepsilon_t &= \sqrt{h_t}\eta_t, \\ h_t &= b_0 + \sum_{j=1}^{q} b_j \varepsilon_{t-j}^2, \end{aligned} \tag{3.1}$$

where $b_0 > 0$, $b_j \geq 0$, $j = 1, \ldots q$, and $\{\eta_t\}$ is a sequence of i.i.d. random variables such that $\mathscr{E}(\eta_t) = 0$, $\mathscr{E}(\eta_t^2) = 1$, following a symmetric distribution, i.e. $\mathscr{E}(\eta_t^3) = 0$, and $\mathscr{E}(\eta_t^4) =: \kappa < \infty$. Let $m_t := h_t(\eta_t^2 - 1)$, then $\mathscr{E}(m_t) = 0$, $\mathscr{E}(m_t^2) = (\kappa - 1)\mathscr{E}(h_t^2)$ and the third equation of (3.1) can be rewritten as

$$\varepsilon_t^2 = h_t + h_t(\eta_t^2 - 1) = b_0 + \sum_{j=1}^{q} b_j \varepsilon_{t-j}^2 + m_t.$$

For simplicity, we write the model in vector and matrix notation as follows:

$$\begin{aligned} X_t &= \mathbf{x_t}' \mathbf{a} + \varepsilon_t, \\ \mathbf{x} &= \mathbf{X} \mathbf{a} + \mathbf{u}, \end{aligned}$$

$$\begin{pmatrix} X_{p+1} \\ X_{p+2} \\ \vdots \\ X_T \end{pmatrix} = \begin{pmatrix} 1 & X_p & X_{p-1} & \cdots & X_1 \\ 1 & X_{p+1} & X_p & \cdots & X_2 \\ \vdots & \vdots & \vdots & \ddots & \vdots \\ 1 & X_{T-1} & X_{T-2} & \cdots & X_{T-p} \end{pmatrix} \begin{pmatrix} a_0 \\ a_1 \\ a_2 \\ \vdots \\ a_p \end{pmatrix} + \begin{pmatrix} \varepsilon_{p+1} \\ \varepsilon_{p+2} \\ \vdots \\ \varepsilon_T \end{pmatrix},$$

$$\begin{aligned} \varepsilon_t^2 &= \mathbf{e_t}' \mathbf{b} + m_t, \\ \mathbf{e} &= \mathbf{E} \mathbf{b} + \mathbf{m}, \end{aligned}$$

$$\begin{pmatrix} \varepsilon_{q+1}^2 \\ \varepsilon_{q+2}^2 \\ \vdots \\ \varepsilon_T^2 \end{pmatrix} = \begin{pmatrix} 1 & \varepsilon_q^2 & \varepsilon_{q-1}^2 & \cdots & \varepsilon_1^2 \\ 1 & \varepsilon_{q+1}^2 & \varepsilon_q^2 & \cdots & \varepsilon_2^2 \\ \vdots & \vdots & \vdots & \ddots & \vdots \\ 1 & \varepsilon_{T-1}^2 & \varepsilon_{T-2}^2 & \cdots & \varepsilon_{T-q}^2 \end{pmatrix} \begin{pmatrix} b_0 \\ b_1 \\ b_2 \\ \vdots \\ b_q \end{pmatrix} + \begin{pmatrix} m_{q+1} \\ m_{q+2} \\ \vdots \\ m_T \end{pmatrix},$$

where $\mathbf{x_t} = (1\ X_{t-1}\ X_{t-2}\ ...\ X_{t-p})'$, $\mathbf{e_t} = (1\ \varepsilon_{t-1}^2\ \varepsilon_{t-2}^2\ ...\ \varepsilon_{t-q}^2)'$.

To prove asymptotic normality of the estimators for the AR part, we make the following assumptions (cf. Nicholls and Pagan (1983), section 2):

(A1) $\{\varepsilon_t,\ t = 1,\ ...,\ T\}$ are martingale differences with

$$\mathscr{E}(\varepsilon_s \varepsilon_t) = 0, \qquad s \neq t,$$
$$= \mathscr{E}(h_t) < \sigma_1^2 < \infty, \qquad s = t.$$

(A2) The roots of the characteristic equation

$$z^p - \sum_{i=1}^{p} a_i z^{p-i} = 0$$

are less than one in absolute value.

(A3) For some positive δ_1

$$\mathscr{E}\left|\mathbf{c_1}'\mathbf{x_t}\varepsilon_t\right|^{2+\delta_1} < \infty,$$

where $\mathbf{c_1}' \in \mathbb{R}^{p+1}$.

(A4) The matrix $\mathbf{A}$ defined

$$\mathbf{A} := \lim_{T\to\infty} \frac{1}{T-p} \sum_{t=p+1}^{T} \mathscr{E}\left(\mathbf{x_t}\mathbf{x_t}'\right)$$

is invertible.

Analogously, we make the following assumptions for the ARCH part:

(B1) $\{m_t,\ t = 1,\ ...,\ T\}$ are martingale differences with

$$\mathscr{E}(m_s m_t) = 0, \qquad s \neq t,$$
$$= (\kappa - 1)\mathscr{E}(h_t^2) < \sigma_2^2 < \infty, \qquad s = t.$$

(B2) The roots of the characteristic equation

$$y^q - \sum_{j=1}^{q} b_j y^{q-j} = 0$$

are less than one in absolute value.

(B3) For some positive δ_2

$$\mathscr{E}\left|\mathbf{c_2}'\mathbf{e_t}m_t\right|^{2+\delta_2} < \infty,$$

where $\mathbf{c_2}' \in \mathbb{R}^{q+1}$.

(B4) The matrix **B** defined

$$\mathbf{B} := \lim_{T\to\infty} \frac{1}{T-q} \sum_{t=q+1}^{T} \mathscr{E}(\mathbf{e_t}\mathbf{e_t}')$$

is invertible.

Remark 3.1
(A2) is equivalent to X_t being causal[1], i.e. X_t can be written in the form

$$X_t = \mu + \sum_{i=0}^{\infty} \alpha_i \varepsilon_{t-i}, \qquad \sum_{i=0}^{\infty} |\alpha_i| < \infty,$$

where

$$\mu = \mathscr{E}(X_t) = \frac{a_0}{1-\sum_{i=1}^{p} a_i}.$$

Remark 3.2
(A1) and (A3) together with Theorem B.1 yield

$$\frac{1}{T-p}\sum_{t=p+1}^{T} \varepsilon_t \xrightarrow{a.s.} \mathscr{E}(\varepsilon_t) \quad and$$

$$\frac{1}{T-p}\sum_{r,\, s=p+1}^{T} \varepsilon_r\varepsilon_s \xrightarrow{a.s.} \mathscr{E}(\varepsilon_r\varepsilon_s).$$

3.1.2 OLS Estimation

In the two-step estimation for the AR(p)-ARCH(q) model it is necessary to estimate the AR part firstly. The ARCH part is then estimated based on the residuals of the AR part. Nicholls and Pagan (1983) proved asymptotic normality of the OLS estimators for the AR(p) model with heteroscedastic errors. Based on their proof now we sketch the asymptotic theory for the AR(p)-ARCH(q) model with the two-step OLS estimation.

[1] See, e.g., Fuller (1996, p.108).

3.1.2.1 AR part

The model to be estimated is

$$X_t = a_0 + \sum_{i=1}^{p} a_i X_{t-i} + \varepsilon_t, \qquad t = p+1, ..., T. \tag{3.2}$$

The OLS estimator $\widehat{\mathbf{a}} = (\widehat{a_0}\ \widehat{a_1}\ ...\ \widehat{a_p})'$ is obtained from the equation

$$\begin{aligned}\widehat{\mathbf{a}} &= \left(\sum_{t=p+1}^{T} \mathbf{x_t}\mathbf{x_t}'\right)\left(\sum_{t=p+1}^{T} \mathbf{x_t}X_t\right) \\ &= \left(\mathbf{X}'\mathbf{X}\right)^{-1}\mathbf{X}'\mathbf{x}.\end{aligned}$$

Note that

$$\begin{aligned}\widehat{\mathbf{a}} &= \left(\mathbf{X}'\mathbf{X}\right)^{-1}\mathbf{X}'\left(\mathbf{X}\mathbf{a}+\mathbf{u}\right) \\ &= \mathbf{a} + \left(\mathbf{X}'\mathbf{X}\right)^{-1}\mathbf{X}'\mathbf{u}\end{aligned}$$

and thus

$$\sqrt{T-p}(\widehat{\mathbf{a}}-\mathbf{a}) = \left(\frac{1}{T-p}\mathbf{X}'\mathbf{X}\right)^{-1}\left(\frac{1}{\sqrt{T-p}}\mathbf{X}'\mathbf{u}\right).$$

Theorem 3.1
Suppose that X_t is generated by the model (3.1) satisfying assumptions (A1)-(A4). Then

$$\sqrt{T-p}(\widehat{\mathbf{a}}-\mathbf{a}) \xrightarrow{d} \mathcal{N}\left(0, \mathbf{A}^{-1}\mathbf{V}\mathbf{A}^{-1}\right),$$

where

$$\mathbf{V} = \lim_{T\to\infty}\frac{1}{T-p}\sum_{t=p+1}^{T}\mathscr{E}\left(h_t\mathbf{x_t}\mathbf{x_t}'\right).$$

Proof. The proof is the mirror image of the Appendix in Nicholls and Pagan (1983). The difference between the model (3.1) and that of Nicholls and Pagan (1983) is that our model contains a constant term a_0 and assumes that h_t is not independent of $\mathbf{x_t}$.

We observe

$$\sqrt{T-p}(\widehat{\mathbf{a}}-\mathbf{a}) = \left(\frac{1}{T-p}\mathbf{X}'\mathbf{X}\right)^{-1}\left(\frac{1}{\sqrt{T-p}}\mathbf{X}'\mathbf{u}\right)$$

and show

$$\frac{1}{T-p}\mathbf{X}'\mathbf{X} \xrightarrow{p} \mathbf{A} \tag{3.3}$$

and

$$\frac{1}{\sqrt{T-p}}\mathbf{X}'\mathbf{u} \xrightarrow{d} \mathcal{N}\left(0, \mathbf{V}\right) \tag{3.4}$$

separately.

Firstly, observe

$$\frac{1}{T-p}\mathbf{X}'\mathbf{X} = \frac{1}{T-p}\sum_{t=p+1}^{T}\begin{pmatrix} 1 & X_{t-1} & X_{t-2} & \dots & X_{t-p} \\ X_{t-1} & X_{t-1}^2 & X_{t-1}X_{t-2} & \dots & X_{t-1}X_{t-p} \\ X_{t-2} & X_{t-2}X_{t-1} & X_{t-2}^2 & \dots & X_{t-2}X_{t-p} \\ \vdots & \vdots & \vdots & \ddots & \vdots \\ X_{t-p} & X_{t-p}X_{t-1} & X_{t-p}X_{t-2} & \dots & X_{t-p}^2 \end{pmatrix}.$$

From Remark 3.1 and 3.2 we obtain

$$\begin{aligned} &\frac{1}{T-p}\sum_{t=p+1}^{T} X_{t-r} \\ =&\frac{1}{T-p}\sum_{t=p+1}^{T}\left(\mu + \sum_{i=0}^{\infty}\alpha_i\varepsilon_{t-r-i}\right) \\ =&\mu + \sum_{i=0}^{\infty}\alpha_i \underbrace{\left(\frac{1}{T-p}\sum_{t=p+1}^{T}\varepsilon_{t-r-i}\right)}_{\xrightarrow{a.s.}\mathscr{E}(\varepsilon_{t-r-i})} \\ &\xrightarrow{a.s.}\mathscr{E}(X_{t-r}) \end{aligned}$$

for $r = 1, \ldots, p$. Analogously, we obtain

$$
\begin{aligned}
&\frac{1}{T-p}\sum_{t=p+1}^{T} X_{t-r}X_{t-s}\\
=&\frac{1}{T-p}\sum_{t=p+1}^{T}\left(\mu+\sum_{i=0}^{\infty}\alpha_i\varepsilon_{t-r-i}\right)\left(\mu+\sum_{j=0}^{\infty}\alpha_j\varepsilon_{t-s-j}\right)\\
=&\frac{1}{T-p}\sum_{t=p+1}^{T}\left(\mu^2+\mu\left(\sum_{i=0}^{\infty}\alpha_i\varepsilon_{t-r-i}+\sum_{j=0}^{\infty}\alpha_j\varepsilon_{t-s-j}\right)\right.\\
&\left.+\sum_{i=0}^{\infty}\sum_{j=0}^{\infty}\alpha_i\alpha_j\varepsilon_{t-r-i}\varepsilon_{t-s-j}\right)\\
=&\mu^2+\mu\left(\sum_{i=0}^{\infty}\alpha_i\underbrace{\frac{1}{T-p}\sum_{t=p+1}^{T}\varepsilon_{t-r-i}}_{\xrightarrow{a.s.}\mathscr{E}(\varepsilon_{t-r-i})}+\sum_{j=0}^{\infty}\alpha_j\underbrace{\frac{1}{T-p}\sum_{t=p+1}^{T}\varepsilon_{t-s-j}}_{\xrightarrow{a.s.}\mathscr{E}(\varepsilon_{t-s-j})}\right)\\
&+\sum_{i=0}^{\infty}\sum_{j=0}^{\infty}\alpha_i\alpha_j\underbrace{\frac{1}{T-p}\sum_{t=p+1}^{T}\varepsilon_{t-r-i}\varepsilon_{t-s-j}}_{\xrightarrow{a.s.}\mathscr{E}(\varepsilon_{t-r-i}\varepsilon_{t-s-j})}\\
&\xrightarrow{a.s.}\mathscr{E}(X_{t-r}X_{t-s})
\end{aligned}
$$

for $r, s = 1, \ldots, p$, therefore,

$$\frac{1}{T-p}\mathbf{X}'\mathbf{X}\xrightarrow{a.s.}\mathbf{A}.$$

Secondly, let $\mathbf{c}\in\mathbb{R}^{p+1}$,

$$\phi_t := \frac{1}{\sqrt{T-p}}\mathbf{c}'\mathbf{x_t}\varepsilon_t, \qquad t = p+1, \ldots, T$$

and $\mathscr{F}_{t-1}$ be the σ-field generated by $\phi_{t-1}, \ldots, \phi_1$. Then

$$\mathscr{E}\left(\phi_t\middle|\mathscr{F}_{t-1}\right)=\frac{1}{\sqrt{T-p}}\mathbf{c}'\mathscr{E}\left(\mathbf{x_t}\sqrt{h_t}\middle|\mathscr{F}_{t-1}\right)\mathscr{E}\left(\eta_t\middle|\mathscr{F}_{t-1}\right)=0$$

and

$$\begin{aligned}
s_T^2 :=& \sum_{t=p+1}^{T} \mathscr{E}(\phi_t^2) \\
=& \frac{1}{T-p} \sum_{t=p+1}^{T} \mathbf{c}' \mathscr{E}\left(\mathbf{x_t x_t}' h_t\right) \mathbf{c} \\
\to& \mathbf{c}'\mathbf{V}\mathbf{c} < \infty \quad as\ T \to \infty,
\end{aligned}$$

that is, ϕ_t is a square integrable martingale difference sequence. Therefore, it suffices to verify the assumptions of a version of the central limit theorem for martingale difference sequences (Brown (1971), see Theorem A.1), namely

$$s_T^{-2} \sum_{t=p+1}^{T} \mathscr{E}\left(\phi_t^2 \middle| \mathscr{F}_{t-1}\right) \xrightarrow{p} 1 \tag{3.5}$$

and for every $\delta > 0$

$$s_T^{-2} \sum_{t=p+1}^{T} \mathscr{E}\left(\phi_t^2 \mathbf{1}\left\{|\phi_t| \geq \delta s_T\right\}\right) \xrightarrow{p} 0. \tag{3.6}$$

On the one hand, (3.5) holds if

$$s_T^2 \left(\sum_{t=p+1}^{T} \mathscr{E}\left(\phi_t^2 \middle| \mathscr{F}_{t-1}\right) - s_T^2 \right) \xrightarrow{p} 0. \tag{3.7}$$

Note that here we already know $s_T^2 \to \mathbf{c}'\mathbf{V}\mathbf{c} < \infty$ as $T \to \infty$. By direct computation we obtain

$$\begin{aligned}
&\sum_{t=p+1}^{T} \mathscr{E}\left(\phi_t^2 \middle| \mathscr{F}_{t-1}\right) - s_T^2 \\
=& \frac{1}{T-p} \sum_{t=p+1}^{T} \mathbf{c}' \left(\mathbf{x_t x_t}' h_t - \mathscr{E}\left(\mathbf{x_t x_t}' h_t\right)\right) \mathbf{c} \\
=& \mathbf{c}' \left(\frac{1}{T-p} \sum_{t=p+1}^{T} \mathbf{x_t x_t}' h_t - \mathscr{E}\left(\mathbf{x_t x_t}' h_t\right)\right) \mathbf{c} \\
\xrightarrow{a.s.}& 0,
\end{aligned}$$

which shows (3.7) and thus (3.5).

On the other hand, from (A3) we obtain the Lyapounov condition (see Corollary A.1)

$$
\begin{aligned}
&\lim_{T\to\infty}\sum_{t=p+1}^{T}\frac{1}{s_T^{2+\delta}}\mathscr{E}\left|\phi_t\right|^{2+\delta}\\
=&\lim_{T\to\infty}\frac{1}{(T-p)^{\delta/2}}\left(\frac{1}{T-p}\sum_{t=p+1}^{T}\mathscr{E}\left|\mathbf{c}'\mathbf{x_t}\varepsilon_t\right|^{2+\delta}\right)\\
=&0
\end{aligned}
$$

for some positive δ, which shows (3.6). The Cramér-Wold theorem and the central limit theorem for martingale difference sequences applied to ϕ_1, ..., ϕ_T show

$$
\frac{1}{\sqrt{T-p}}\mathbf{X}'\mathbf{u}\xrightarrow{d}\mathscr{N}\left(0,\mathbf{V}\right)
$$

and the Slutsky theorem yields the desired result. □

3.1.2.2 ARCH part

3.1.2.2.1 The imaginary case where ε_t are known

Firstly, we analyse the imaginary case where $\{\varepsilon_t,\ t=1,\ ...,\ T\}$ are known. The model to be estimated is then

$$
\varepsilon_t^2=b_0+\sum_{j=1}^{q}b_j\varepsilon_{t-j}^2+m_t,\qquad t=q+1,...,T. \tag{3.8}
$$

The OLS estimator $\widehat{\mathbf{b}}=(\widehat{b_0}\ \widehat{b_1}\ ...\ \widehat{b_q})'$ is obtained from the equation

$$
\begin{aligned}
\widehat{\mathbf{b}}=&\left(\sum_{t=q+1}^{T}\mathbf{e_t}\mathbf{e_t}'\right)\left(\sum_{t=q+1}^{T}\mathbf{e_t}\varepsilon_t^2\right)\\
=&\left(\mathbf{E}'\mathbf{E}\right)^{-1}\mathbf{E}'\mathbf{e}.
\end{aligned}
$$

Note that

$$
\begin{aligned}
\widehat{\mathbf{b}}=&\left(\mathbf{E}'\mathbf{E}\right)^{-1}\mathbf{E}'\left(\mathbf{E}\mathbf{b}+\mathbf{m}\right)\\
=&\mathbf{b}+\left(\mathbf{E}'\mathbf{E}\right)^{-1}\mathbf{E}'\mathbf{m}
\end{aligned}
$$

and thus

$$\sqrt{T-q}(\widehat{\mathbf{b}}-\mathbf{b}) = \left(\frac{1}{T-q}\mathbf{E}'\mathbf{E}\right)^{-1}\left(\frac{1}{\sqrt{T-q}}\mathbf{E}'\mathbf{m}\right).$$

Using arguments analogous to Theorem 3.1, we obtain the following corollary.

Corollary 3.1
Suppose that ε_t^2 is generated by the model (3.8) satisfying assumptions (B1)-(B4). Then

$$\sqrt{T-q}(\widehat{\mathbf{b}}-\mathbf{b}) \xrightarrow{d} \mathcal{N}\left(0, \mathbf{B}^{-1}\mathbf{W}\mathbf{B}^{-1}\right),$$

where

$$\mathbf{W} = \lim_{T\to\infty} \frac{\kappa-1}{T-q} \sum_{t=q+1}^{T} \mathscr{E}(h_t^2 \mathbf{e_t}\mathbf{e_t}').$$

3.1.2.2.2 The standard case where ε_t are unknown

Secondly, we analyse the standard case where $\{\varepsilon_t,\ t=1,\ ...,\ T\}$ are unknown. In this case we adopt the two-step estimation, where the residuals of the AR part

$$\widehat{\varepsilon_t} = X_t - \widehat{a_0} - \sum_{i=1}^{p} \widehat{a_i} X_{t-i}, \quad t = p+1,\ ...,\ T$$

are used to estimate the ARCH part. The model to be estimated is now

$$\widehat{\varepsilon_t}^2 = b_0 + \sum_{j=1}^{q} b_j \widehat{\varepsilon_{t-j}}^2 + n_t, \quad t = p+q+1,\ ...,\ T, \tag{3.9}$$

or in vector and matrix notation

$$\begin{aligned}
\widehat{\varepsilon_t}^2 &= \widehat{\mathbf{e_t}}' \mathbf{b} + n_t,\\
\widehat{\mathbf{e}} &= \widehat{\mathbf{E}} \mathbf{b} + \mathbf{n},
\end{aligned}$$

$$\begin{pmatrix} \widehat{\varepsilon_{p+q+1}}^2 \\ \widehat{\varepsilon_{p+q+2}}^2 \\ \vdots \\ \widehat{\varepsilon_T}^2 \end{pmatrix} = \begin{pmatrix} 1 & \widehat{\varepsilon_{p+q}}^2 & \widehat{\varepsilon_{p+q-1}}^2 & \cdots & \widehat{\varepsilon_1}^2 \\ 1 & \widehat{\varepsilon_{p+q+1}}^2 & \widehat{\varepsilon_{p+q}}^2 & \cdots & \widehat{\varepsilon_2}^2 \\ \vdots & \vdots & \vdots & \ddots & \vdots \\ 1 & \widehat{\varepsilon_{T-1}}^2 & \widehat{\varepsilon_{T-2}}^2 & \cdots & \widehat{\varepsilon_{T-q}}^2 \end{pmatrix} \begin{pmatrix} b_0 \\ b_1 \\ b_2 \\ \vdots \\ b_q \end{pmatrix} + \begin{pmatrix} n_{p+q+1} \\ n_{p+q+2} \\ \vdots \\ n_T \end{pmatrix},$$

where $\widehat{\mathbf{e_t}} = (1\ \widehat{\varepsilon_{t-1}}^2\ \widehat{\varepsilon_{t-2}}^2\ ...\ \widehat{\varepsilon_{t-q}}^2)'$.

The OLS estimator $\widetilde{\mathbf{b}} = (\widetilde{b_0}\ \widetilde{b_1}\ ...\widetilde{b_q})'$ is obtained from the equation

$$\widetilde{\mathbf{b}} = \left(\sum_{t=p+q+1}^{T} \widehat{\mathbf{e}}_t\widehat{\mathbf{e}}_t'\right)\left(\sum_{t=p+q+1}^{T} \widehat{\mathbf{e}}_t\widehat{\varepsilon}_t^{\,2}\right)$$
$$= \left(\widehat{\mathbf{E}}'\widehat{\mathbf{E}}\right)^{-1}\widehat{\mathbf{E}}'\widehat{\mathbf{e}}.$$

Note that

$$\widetilde{\mathbf{b}} = \left(\widehat{\mathbf{E}}'\widehat{\mathbf{E}}\right)^{-1}\widehat{\mathbf{E}}'\left(\widehat{\mathbf{E}}\mathbf{b}+\mathbf{n}\right)$$
$$= \mathbf{b} + \left(\widehat{\mathbf{E}}'\widehat{\mathbf{E}}\right)^{-1}\widehat{\mathbf{E}}'\mathbf{n}$$

and thus

$$\sqrt{T-p-q}(\widetilde{\mathbf{b}}-\mathbf{b}) = \left(\frac{1}{T-p-q}\widehat{\mathbf{E}}'\widehat{\mathbf{E}}\right)^{-1}\left(\frac{1}{\sqrt{T-p-q}}\widehat{\mathbf{E}}'\mathbf{n}\right).$$

Remark 3.3

For simplicity, we denote

$$\varepsilon_t - \widehat{\varepsilon}_t = \mathbf{x_t}'(\widehat{\mathbf{a}}-\mathbf{a}) = \sum_{k=0}^{p} x_{t,k}(\widehat{a_k}-a_k),$$

where

$$\mathbf{x_t} = (1\ X_{t-1}\ X_{t-2}\ ...\ X_{t-p})' = (x_{t,0}\ x_{t,1}\ x_{t,2}\ ...\ x_{t,p})',$$
$$\mathbf{a} = (a_0\ a_1\ ...\ a_p)' \quad \textit{and} \quad \widehat{\mathbf{a}} = (\widehat{a_0}\ \widehat{a_1}\ ...\ \widehat{a_p})'$$

throughout this chapter without further notice.

In fact, the estimator $\widetilde{\mathbf{b}}$ is asymptotically as efficient as $\widehat{\mathbf{b}}$ in §3.1.2.2.1, as we will see in the following theorem.

Theorem 3.2

Suppose that X_t is generated by the model (3.1) satisfying assumptions (A1)-(A4) and (B1)-(B4). Then

$$\sqrt{T-p-q}(\widetilde{\mathbf{b}}-\mathbf{b}) \xrightarrow{d} \mathscr{N}\left(0, \mathbf{B}^{-1}\mathbf{W}\mathbf{B}^{-1}\right).$$

Proof. From Corollary 3.1 we obtain

$$\frac{1}{T-p-q}\sum_{t=p+q+1}^{T} \mathbf{e_t}\mathbf{e_t}' \xrightarrow{p} \mathbf{B},$$
$$\frac{1}{\sqrt{T-p-q}}\sum_{t=p+q+1}^{T} \mathbf{e_t}m_t \xrightarrow{d} \mathscr{N}\left(0, \mathbf{W}\right).$$

Together with the Slutsky theorem it suffices to show

$$\frac{1}{T-p-q}\sum_{t=p+q+1}^{T}\widehat{\mathbf{e_t}}\widehat{\mathbf{e_t}}' = \frac{1}{T-p-q}\sum_{t=p+q+1}^{T}\mathbf{e_t}\mathbf{e_t}' + o_p(1), \tag{3.10}$$

$$\frac{1}{\sqrt{T-p-q}}\sum_{t=p+q+1}^{T}\widehat{\mathbf{e_t}}n_t = \frac{1}{\sqrt{T-p-q}}\sum_{t=p+q+1}^{T}\mathbf{e_t}m_t + o_p(1). \tag{3.11}$$

To prove (3.10), it is sufficient to show the following properties for all $i, j \in \mathbb{Z}$:

$$\frac{1}{\sqrt{T}}\sum_{t=1}^{T}\left(\widehat{\varepsilon_{t-i}}^2 - \varepsilon_{t-i}^2\right) = o_p(1), \tag{3.12}$$

$$\frac{1}{\sqrt{T}}\sum_{t=1}^{T}\varepsilon_{t-i}^2\left(\widehat{\varepsilon_{t-j}}^2 - \varepsilon_{t-j}^2\right) = o_p(1), \tag{3.13}$$

$$\frac{1}{\sqrt{T}}\sum_{t=1}^{T}\widehat{\varepsilon_{t-i}}^2\left(\widehat{\varepsilon_{t-j}}^2 - \varepsilon_{t-j}^2\right) = o_p(1), \tag{3.14}$$

$$\frac{1}{\sqrt{T}}\sum_{t=1}^{T}\left(\widehat{\varepsilon_{t-i}}^2\widehat{\varepsilon_{t-j}}^2 - \varepsilon_{t-i}^2\varepsilon_{t-j}^2\right) = o_p(1). \tag{3.15}$$

Firstly, the left hand side of (3.12) equals

$$\begin{aligned}
&\frac{1}{\sqrt{T}}\sum_{t=1}^{T}\left(\widehat{\varepsilon_{t-i}}^2 - \varepsilon_{t-i}^2\right)\\
=&\frac{1}{\sqrt{T}}\sum_{t=1}^{T}\left(\widehat{\varepsilon_{t-i}} - \varepsilon_{t-i}\right)^2 + \frac{2}{\sqrt{T}}\sum_{t=1}^{T}\varepsilon_{t-i}\left(\widehat{\varepsilon_{t-i}} - \varepsilon_{t-i}\right)\\
=&\frac{1}{\sqrt{T}}\sum_{t=1}^{T}\left(\mathbf{x_{t-i}}'(\mathbf{a}-\widehat{\mathbf{a}})\right)^2 + \frac{2}{\sqrt{T}}\sum_{t=1}^{T}\varepsilon_{t-i}\mathbf{x_{t-i}}'(\mathbf{a}-\widehat{\mathbf{a}})\\
\le&\frac{1}{\sqrt{T}}\sum_{t=1}^{T}\left(\sum_{k=0}^{p}x_{t-i,k}^2\sum_{k=0}^{p}(a_k-\widehat{a_k})^2\right) + \underbrace{\left(\frac{2}{T}\sum_{t=1}^{T}\varepsilon_{t-i}\mathbf{x_{t-i}}'\right)}_{=o_p(1)}\underbrace{\sqrt{T}(\mathbf{a}-\widehat{\mathbf{a}})}_{=O_p(1)}\\
=&\underbrace{\sqrt{T}\sum_{k=0}^{p}(a_k-\widehat{a_k})^2}_{=O_p(T^{-1/2})}\frac{1}{T}\sum_{t=1}^{T}\underbrace{\sum_{k=0}^{p}x_{t-i,k}^2}_{=O_p(1)} + o_p(1)\\
=&o_p(1).
\end{aligned}$$

Here we obtain

$$\mathbf{z_{t-i}}' := \frac{2}{T}\sum_{t=1}^{T} \varepsilon_{t-i}\mathbf{x_{t-i}}' = o_p(1)$$

because $\mathscr{E}\left(\mathbf{z_{t-i}}'\right) = 0$,

$$\mathscr{E}\left(\mathbf{z_{t-i}}'\mathbf{z_{t-i}}\right) = \frac{4}{T}\left(\frac{1}{T}\sum_{t=1}^{T}\mathscr{E}\left(\varepsilon_{t-i}^2\mathbf{x_{t-i}}'\mathbf{x_{t-i}}\right)\right) = o(1),$$

and, therefore, from Chebyshev's inequality we obtain for every $\delta > 0$

$$P\left\{\left|\mathbf{z_{t-i}}' - \mathscr{E}\left(\mathbf{z_{t-i}}'\right)\right| > \delta\right\} \leq \frac{1}{\delta^2}\mathscr{E}\left(\mathbf{z_{t-i}}'\mathbf{z_{t-i}}\right) = o(1),$$

that is, according to the definition

$$\mathbf{z_{t-i}}' \xrightarrow{p} \mathscr{E}\left(\mathbf{z_{t-i}}'\right) = 0.$$

Secondly, the proof of (3.13) is the mirror image of (3.12).

$$\begin{aligned}
&\frac{1}{\sqrt{T}}\sum_{t=1}^{T}\varepsilon_{t-i}^2\left(\widehat{\varepsilon_{t-j}}^2 - \varepsilon_{t-j}^2\right)\\
=&\frac{1}{\sqrt{T}}\sum_{t=1}^{T}\varepsilon_{t-i}^2\left(\widehat{\varepsilon_{t-j}} - \varepsilon_{t-j}\right)^2 + \frac{2}{\sqrt{T}}\sum_{t=1}^{T}\varepsilon_{t-i}^2\varepsilon_{t-j}\left(\widehat{\varepsilon_{t-j}} - \varepsilon_{t-j}\right)\\
=&\frac{1}{\sqrt{T}}\sum_{t=1}^{T}\varepsilon_{t-i}^2\left(\mathbf{x_{t-j}}'(\mathbf{a}-\widehat{\mathbf{a}})\right)^2 + \frac{2}{\sqrt{T}}\sum_{t=1}^{T}\varepsilon_{t-i}^2\varepsilon_{t-j}\mathbf{x_{t-j}}'(\mathbf{a}-\widehat{\mathbf{a}})\\
\leq&\frac{1}{\sqrt{T}}\sum_{t=1}^{T}\varepsilon_{t-i}^2\left(\sum_{k=0}^{p}x_{t-j,k}^2\sum_{k=0}^{p}(a_k-\widehat{a_k})^2\right) + \underbrace{\left(\frac{2}{T}\sum_{t=1}^{T}\varepsilon_{t-i}^2\varepsilon_{t-j}\mathbf{x_{t-j}}'\right)}_{=o_p(1)}\underbrace{\sqrt{T}(\mathbf{a}-\widehat{\mathbf{a}})}_{=O_p(1)}\\
=&\underbrace{\sqrt{T}\sum_{k=0}^{p}(a_k-\widehat{a_k})^2}_{=O_p(T^{-1/2})}\frac{1}{T}\sum_{t=1}^{T}\varepsilon_{t-i}^2\underbrace{\sum_{k=0}^{p}x_{t-j,k}^2}_{=O_p(1)} + o_p(1)\\
=&o_p(1).
\end{aligned}$$

Thirdly, the left hand side of (3.14) equals

$$
\begin{aligned}
&\frac{1}{\sqrt{T}}\sum_{t=1}^{T}\widehat{\varepsilon_{t-i}}^{2}\left(\widehat{\varepsilon_{t-j}}^{2}-\varepsilon_{t-j}^{2}\right)\\
=&\frac{1}{\sqrt{T}}\sum_{t=1}^{T}\left(\widehat{\varepsilon_{t-j}}^{2}-\varepsilon_{t-j}^{2}\right)\left(\left(\widehat{\varepsilon_{t-i}}^{2}-\varepsilon_{t-i}^{2}\right)+\varepsilon_{t-i}^{2}\right)\\
=&\frac{1}{\sqrt{T}}\sum_{t=1}^{T}\left\{\left(\widehat{\varepsilon_{t-j}}-\varepsilon_{t-j}\right)^{2}+2\varepsilon_{t-j}\left(\widehat{\varepsilon_{t-j}}-\varepsilon_{t-j}\right)\right\}\\
&\left\{\left(\left(\widehat{\varepsilon_{t-i}}-\varepsilon_{t-i}\right)^{2}+2\varepsilon_{t-i}\left(\widehat{\varepsilon_{t-i}}-\varepsilon_{t-i}\right)\right)+\varepsilon_{t-i}^{2}\right\}\\
=&\underbrace{\frac{1}{\sqrt{T}}\sum_{t=1}^{T}\left(\widehat{\varepsilon_{t-j}}-\varepsilon_{t-j}\right)^{2}\left(\widehat{\varepsilon_{t-i}}-\varepsilon_{t-i}\right)^{2}}_{=:s_1}\\
&+\underbrace{\frac{2}{\sqrt{T}}\sum_{t=1}^{T}\varepsilon_{t-i}\left(\widehat{\varepsilon_{t-j}}-\varepsilon_{t-j}\right)^{2}\left(\widehat{\varepsilon_{t-i}}-\varepsilon_{t-i}\right)}_{=:s_2}\\
&+\frac{1}{\sqrt{T}}\sum_{t=1}^{T}\varepsilon_{t-i}^{2}\left(\widehat{\varepsilon_{t-j}}-\varepsilon_{t-j}\right)^{2}\\
&+\underbrace{\frac{2}{\sqrt{T}}\sum_{t=1}^{T}\varepsilon_{t-j}\left(\widehat{\varepsilon_{t-j}}-\varepsilon_{t-j}\right)\left(\widehat{\varepsilon_{t-i}}-\varepsilon_{t-i}\right)^{2}}_{=:s_3}\\
&+\underbrace{\frac{4}{\sqrt{T}}\sum_{t=1}^{T}\varepsilon_{t-j}\varepsilon_{t-i}\left(\widehat{\varepsilon_{t-j}}-\varepsilon_{t-j}\right)\left(\widehat{\varepsilon_{t-i}}-\varepsilon_{t-i}\right)}_{=:s_4}\\
&+\frac{2}{\sqrt{T}}\sum_{t=1}^{T}\varepsilon_{t-j}\varepsilon_{t-i}^{2}\left(\widehat{\varepsilon_{t-j}}-\varepsilon_{t-j}\right)\\
\overset{(3.13)}{=}&s_1+s_2+s_3+s_4+o_p(1).
\end{aligned}
$$

We prove now that s_1, s_2, s_3 and s_4 are $o_p(1)$ respectively.

If i≠j

$$
\begin{aligned}
s_1 =& \frac{1}{\sqrt{T}} \sum_{t=1}^{T} \left(\mathbf{x_{t-j}}'(\mathbf{a}-\widehat{\mathbf{a}})\right)^2 \left(\mathbf{x_{t-i}}'(\mathbf{a}-\widehat{\mathbf{a}})\right)^2 \\
\leq& \frac{1}{\sqrt{T}} \sum_{t=1}^{T} \left(\sum_{k=0}^{p} x_{t-j,k}^2 \sum_{k=0}^{p} (a_k-\widehat{a_k})^2 \sum_{k=0}^{p} x_{t-i,k}^2 \sum_{k=0}^{p} (a_k-\widehat{a_k})^2 \right) \\
=& \underbrace{\sqrt{T}\left(\sum_{k=0}^{p}(a_k-\widehat{a_k})^2\right)^2}_{=O_p(T^{-3/2})} \frac{1}{T}\sum_{t=1}^{T} \underbrace{\left(\sum_{k=0}^{p} x_{t-j,k}^2 \sum_{k=0}^{p} x_{t-i,k}^2\right)}_{=O_p(1)} \\
=& o_p(1).
\end{aligned}
$$

$$
\begin{aligned}
s_2 =& \frac{2}{\sqrt{T}} \sum_{t=1}^{T} \varepsilon_{t-i} \left(\mathbf{x_{t-j}}'(\mathbf{a}-\widehat{\mathbf{a}})\right)^2 \mathbf{x_{t-i}}'(\mathbf{a}-\widehat{\mathbf{a}}) \\
\leq& \frac{2}{T} \sum_{t=1}^{T} \left(\varepsilon_{t-i} \sum_{k=0}^{p} x_{t-j,k}^2 \sum_{k=0}^{p} (a_k-\widehat{a_k})^2 \mathbf{x_{t-i}}' \right) \sqrt{T}(\mathbf{a}-\widehat{\mathbf{a}}) \\
=& 2\underbrace{\sum_{k=0}^{p}(a_k-\widehat{a_k})^2}_{=O_p(T^{-1})} \frac{1}{T}\sum_{t=1}^{T} \underbrace{\left(\varepsilon_{t-i}\sum_{k=0}^{p} x_{t-j,k}^2 \mathbf{x_{t-i}}'\right)}_{=O_p(1)} \underbrace{\sqrt{T}(\mathbf{a}-\widehat{\mathbf{a}})}_{=O_p(1)} \\
=& o_p(1).
\end{aligned}
$$

$$
s_3 = 2\underbrace{\sum_{k=0}^{p}(a_k-\widehat{a_k})^2}_{=O_p(T^{-1})} \frac{1}{T}\sum_{t=1}^{T} \underbrace{\left(\varepsilon_{t-j}\sum_{k=0}^{p} x_{t-i,\,k}^2 \mathbf{x_{t-j}}'\right)}_{=O_p(1)} \underbrace{\sqrt{T}(\mathbf{a}-\widehat{\mathbf{a}})}_{=O_p(1)} = o_p(1).
$$

$$
\begin{aligned}
s_4 =& \frac{4}{\sqrt{T}} \sum_{t=1}^{T} \varepsilon_{t-j}\varepsilon_{t-i}\mathbf{x_{t-j}}'(\mathbf{a}-\widehat{\mathbf{a}})\,\mathbf{x_{t-i}}'(\mathbf{a}-\widehat{\mathbf{a}}) \\
\leq& \frac{4}{\sqrt{T}} \sum_{t=1}^{T} |\varepsilon_{t-j}\varepsilon_{t-i}| \left|\mathbf{x_{t-j}}'(\mathbf{a}-\widehat{\mathbf{a}})\right| \left|\mathbf{x_{t-i}}'(\mathbf{a}-\widehat{\mathbf{a}})\right| \\
\leq& \frac{4}{\sqrt{T}} \sum_{t=1}^{T} |\varepsilon_{t-j}\varepsilon_{t-i}| \sqrt{\sum_{k=0}^{p} x_{t-j,k}^2} \sqrt{\sum_{k=0}^{p} (a_k-\widehat{a_k})^2} \sqrt{\sum_{k=0}^{p} x_{t-i,k}^2} \sqrt{\sum_{k=0}^{p} (a_k-\widehat{a_k})^2} \\
=& \underbrace{4\sqrt{T} \sum_{k=0}^{p} (a_k-\widehat{a_k})^2}_{=O_p(T^{-1/2})} \frac{1}{T} \sum_{t=1}^{T} \underbrace{|\varepsilon_{t-j}\varepsilon_{t-i}| \sqrt{\sum_{k=0}^{p} x_{t-j,k}^2} \sqrt{\sum_{k=0}^{p} x_{t-i,k}^2}}_{=O_p(1)} \\
=& o_p(1).
\end{aligned}
$$

If i=j

$$
\begin{aligned}
s_1 =& \frac{1}{\sqrt{T}} \sum_{t=1}^{T} \left(\mathbf{x_{t-i}}'(\mathbf{a}-\widehat{\mathbf{a}})\right)^4 \\
\leq& \frac{1}{\sqrt{T}} \sum_{t=1}^{T} \left(\sum_{k=0}^{p} x_{t-i,k}^2 \sum_{k=0}^{p} (a_k-\widehat{a_k})^2 \right)^2 \\
=& \underbrace{\sqrt{T} \left(\sum_{k=0}^{p} (a_k-\widehat{a_k})^2 \right)^2}_{=O_p(T^{-3/2})} \frac{1}{T} \sum_{t=1}^{T} \underbrace{\left(\sum_{k=0}^{p} x_{t-i,k}^2 \right)^2}_{=O_p(1)} \\
=& o_p(1).
\end{aligned}
$$

$$\begin{aligned}
s_2 = s_3 =& \frac{2}{\sqrt{T}} \sum_{t=1}^{T} \varepsilon_{t-i} \left(\mathbf{x_{t-i}}' (\mathbf{a} - \widehat{\mathbf{a}}) \right)^3 \\
\leq & \frac{2}{\sqrt{T}} \sum_{t=1}^{T} |\varepsilon_{t-i}| \left| \mathbf{x_{t-i}}' (\mathbf{a} - \widehat{\mathbf{a}}) \right|^3 \\
\leq & \frac{2}{\sqrt{T}} \sum_{t=1}^{T} |\varepsilon_{t-i}| \left(\sum_{k=0}^{p} x_{t-i,k}^2 \right)^{3/2} \left(\sum_{k=0}^{p} (a_k - \widehat{a_k})^2 \right)^{3/2} \\
=& \underbrace{2\sqrt{T} \left(\sum_{k=0}^{p} (a_k - \widehat{a_k})^2 \right)^{3/2}}_{=O_p(T^{-1})} \frac{1}{T} \sum_{t=1}^{T} \underbrace{|\varepsilon_{t-i}| \left(\sum_{k=0}^{p} x_{t-i,k}^2 \right)^{3/2}}_{=O_p(1)} \\
=& o_p(1).
\end{aligned}$$

$$\begin{aligned}
s_4 =& \frac{4}{\sqrt{T}} \sum_{t=1}^{T} \varepsilon_{t-i}^2 \left(\mathbf{x_{t-i}}' (\mathbf{a} - \widehat{\mathbf{a}}) \right)^2 \\
\leq & \frac{4}{\sqrt{T}} \sum_{t=1}^{T} \varepsilon_{t-i}^2 \left(\sum_{k=0}^{p} x_{t-i,k}^2 \sum_{k=0}^{p} (a_k - \widehat{a_k})^2 \right) \\
=& \underbrace{4\sqrt{T} \sum_{k=0}^{p} (a_k - \widehat{a_k})^2}_{=O_p(T^{-1/2})} \frac{1}{T} \sum_{t=1}^{T} \underbrace{\varepsilon_{t-i}^2 \sum_{k=0}^{p} x_{t-i,k}^2}_{=O_p(1)} \\
=& o_p(1).
\end{aligned}$$

Finally, from (3.13) and (3.14) we obtain (3.15).

$$\frac{1}{\sqrt{T}} \sum_{t=1}^{T} \widehat{\varepsilon_{t-i}}^2 \widehat{\varepsilon_{t-j}}^2 \overset{(3.14)}{=} \frac{1}{\sqrt{T}} \sum_{t=1}^{T} \widehat{\varepsilon_{t-i}}^2 \varepsilon_{t-j}^2 + o_p(1) \overset{(3.13)}{=} \frac{1}{\sqrt{T}} \sum_{t=1}^{T} \varepsilon_{t-i}^2 \varepsilon_{t-j}^2 + o_p(1).$$

To prove (3.11), note that

$$\begin{aligned}
m_t &= \varepsilon_t^2 - b_0 - \sum_{j=1}^{q} b_j \varepsilon_{t-j}^2, \\
n_t &= \widehat{\varepsilon_t}^2 - b_0 - \sum_{j=1}^{q} b_j \widehat{\varepsilon_{t-j}}^2 \\
&= m_t + \left(\widehat{\varepsilon_t}^2 - \varepsilon_t^2 \right) - \sum_{j=1}^{q} b_j \left(\widehat{\varepsilon_{t-j}}^2 - \varepsilon_{t-j}^2 \right),
\end{aligned}$$

and thus we obtain for $i = 1, ..., q$

$$\begin{aligned}
&\frac{1}{\sqrt{T-p-q}} \sum_{t=p+q+1}^{T} n_t \widehat{\varepsilon_{t-i}}^2 \\
=&\frac{1}{\sqrt{T-p-q}} \sum_{t=p+q+1}^{T} m_t \widehat{\varepsilon_{t-i}}^2 + \underbrace{\frac{1}{\sqrt{T-p-q}} \sum_{t=p+q+1}^{T} \left(\widehat{\varepsilon_t}^2 - \varepsilon_t^2\right) \widehat{\varepsilon_{t-i}}^2}_{\overset{(3.14)}{=} o_p(1)} \\
&- \sum_{j=1}^{q} b_j \underbrace{\frac{1}{\sqrt{T-p-q}} \sum_{t=p+q+1}^{T} \left(\widehat{\varepsilon_{t-j}}^2 - \varepsilon_{t-j}^2\right) \widehat{\varepsilon_{t-i}}^2}_{\overset{(3.14)}{=} o_p(1)} \\
=&\frac{1}{\sqrt{T-p-q}} \sum_{t=p+q+1}^{T} m_t \widehat{\varepsilon_{t-i}}^2 + o_p(1).
\end{aligned}$$

Therefore, to prove (3.11), it suffices to show for $i = 1, ..., q$

$$\frac{1}{\sqrt{T-p-q}} \sum_{t=p+q+1}^{T} m_t \left(\widehat{\varepsilon_{t-i}}^2 - \varepsilon_{t-i}^2\right) = o_p(1).$$

Analogously to the proof of (3.13), we obtain

$$\begin{aligned}
&\frac{1}{\sqrt{T}} \sum_{t=1}^{T} m_t \left(\widehat{\varepsilon_{t-i}}^2 - \varepsilon_{t-i}^2\right) \\
\leq&\frac{1}{\sqrt{T}} \sum_{t=1}^{T} m_t \left(\sum_{k=0}^{p} x_{t-i,k}^2 \sum_{k=0}^{p} (a_k - \widehat{a_k})^2\right) + \underbrace{\left(\frac{2}{T} \sum_{t=1}^{T} m_t \varepsilon_{t-i} \mathbf{x_{t-i}}'\right)}_{=o_p(1)} \underbrace{\sqrt{T}\left(\mathbf{a} - \widehat{\mathbf{a}}\right)}_{=O_p(1)} \\
=&\underbrace{\sqrt{T} \sum_{k=0}^{p} (a_k - \widehat{a_k})^2}_{=O_p(T^{-1/2})} \frac{1}{T} \sum_{t=1}^{T} m_t \underbrace{\sum_{k=0}^{p} x_{t-i,k}^2}_{=O_p(1)} + o_p(1) \\
=&o_p(1).
\end{aligned}$$

□

3.2 Residual Bootstrap

Kreiß (1997) proved the asymptotic validity of the residual bootstrap technique applied to the Yule-Walker estimators for the AR(p) model with homoscedastic errors. It is also shown that the wild bootstrap technique applied to the OLS estimators for the AR(p) model with heteroscedastic errors is weakly consistent. In this and the following section we introduce possible ways to bootstrap the AR(p)-ARCH(q) model based on Kreiß (1997).

In this section a residual bootstrap technique is proposed and its consistency proved. A residual bootstrap method can be applied to the model (3.1) as follows:

(step 1) Obtain the OLS estimator $\widehat{\mathbf{a}}$ and calculate the residuals

$$\widehat{\varepsilon_t} = X_t - \widehat{a_0} - \sum_{i=1}^{p} \widehat{a_i} X_{t-i}, \qquad t = p+1, \ldots, T.$$

(step 2) Compute the OLS estimator $\widetilde{\mathbf{b}}$ and calculate the estimated heteroscedasticity

$$\widetilde{h_t} = \widetilde{b_0} + \sum_{j=1}^{q} \widetilde{b_j} \widehat{\varepsilon_{t-j}}^2, \qquad t = p+q+1, \ldots, T.$$

(step 3) Compute the estimated bias

$$\widehat{\eta_t} = \frac{\widehat{\varepsilon_t}}{\sqrt{\widetilde{h_t}}}$$

for $t = p+q+1, \ldots, T$ and the standardised estimated bias

$$\widetilde{\eta_t} = \frac{\widehat{\eta_t} - \widehat{\mu}}{\widehat{\sigma}}$$

for $t = p+q+1, \ldots, T$, where

$$\widehat{\mu} = \frac{1}{T-p-q} \sum_{t=p+q+1}^{T} \widehat{\eta_t} \quad \textit{and} \quad \widehat{\sigma}^2 = \frac{1}{T-p-q} \sum_{t=p+q+1}^{T} (\widehat{\eta_t} - \widehat{\mu})^2.$$

(step 4) Obtain the empirical distribution function $\mathscr{F}_T(x)$ based on $\widetilde{\eta_t}$ defined by

$$\mathscr{F}_T(x) := \frac{1}{T-p-q} \sum_{t=p+q+1}^{T} \mathbf{1}(\widetilde{\eta_t} \leq x).$$

(step 5) Generate the bootstrap process X_t^* by computing

$$X_t^* = \widehat{a}_0 + \sum_{i=1}^{p} \widehat{a}_i X_{t-i} + \varepsilon_t^*$$

$$\varepsilon_t^* = \sqrt{\widetilde{h}_t}\eta_t^*, \quad \eta_t^* \overset{iid}{\sim} \mathscr{F}_T(x), \quad t = p+q+1, \ldots, T,$$

or in vector and matrix notation

$$\begin{array}{ccccc} \mathbf{x}^* & = & \mathbf{X} & \widehat{\mathbf{a}} & + \quad \mathbf{u}^*, \end{array}$$

$$\begin{pmatrix} X_{p+q+1}^* \\ X_{p+q+2}^* \\ \vdots \\ X_T^* \end{pmatrix} = \begin{pmatrix} 1 & X_{p+q} & X_{p+q-1} & \ldots & X_{q+1} \\ 1 & X_{p+q+1} & X_{p+q} & \ldots & X_{q+2} \\ \vdots & \vdots & \vdots & \ddots & \vdots \\ 1 & X_{T-1} & X_{T-2} & \ldots & X_{T-p} \end{pmatrix} \begin{pmatrix} \widehat{a}_0 \\ \widehat{a}_1 \\ \widehat{a}_2 \\ \vdots \\ \widehat{a}_p \end{pmatrix} + \begin{pmatrix} \varepsilon_{p+q+1}^* \\ \varepsilon_{p+q+2}^* \\ \vdots \\ \varepsilon_T^* \end{pmatrix}.$$

(step 6) Calculate the bootstrap estimator

$$\widehat{\mathbf{a}^*} = (\mathbf{X}'\mathbf{X})^{-1}\mathbf{X}'\mathbf{x}^*,$$
$$\widetilde{\mathbf{b}^*} = (\widehat{\mathbf{E}}'\widehat{\mathbf{E}})^{-1}\widehat{\mathbf{E}}'\mathbf{e}^*,$$

where $\mathbf{e}^* = \left(\varepsilon_{p+q+1}^{*2} \ \varepsilon_{p+q+2}^{*2} \cdots \varepsilon_T^{*2}\right)'$.

Remark 3.4

η_t^* has the following properties:

$$\begin{aligned} \mathscr{E}_*(\eta_t^*) &= \sum_{t=1}^{T} \left(\frac{\widehat{\eta}_t - \widehat{\mu}}{\widehat{\sigma}}\right)\frac{1}{T} \\ &= \frac{1}{\widehat{\sigma}}\Big(\underbrace{\frac{1}{T}\sum_{t=1}^{T}\widehat{\eta}_t}_{=\widehat{\mu}} - \widehat{\mu}\Big) \\ &= 0, \end{aligned}$$

$$\begin{aligned} \mathscr{E}_*(\eta_t^{*2}) &= \sum_{t=1}^{T} \left(\frac{\widehat{\eta}_t - \widehat{\mu}}{\widehat{\sigma}}\right)^2 \frac{1}{T} \\ &= \frac{1}{\widehat{\sigma}^2} \underbrace{\frac{1}{T}\sum_{t=1}^{T}(\widehat{\eta}_t - \widehat{\mu})^2}_{=\widehat{\sigma}^2} \\ &= 1, \end{aligned}$$

$$\begin{aligned}\mathscr{E}_*(\eta_t^{*3}) &= \sum_{t=1}^{T}\left(\frac{\widehat{\eta}_t-\widehat{\mu}}{\widehat{\sigma}}\right)^3\frac{1}{T}\\ &= \frac{1}{\widehat{\sigma}^3}\frac{1}{T}\sum_{t=1}^{T}(\widehat{\eta}_t-\widehat{\mu})^3\\ &= o_p(1),\end{aligned}$$

$$\begin{aligned}\mathscr{E}_*(\eta_t^{*4}) &= \sum_{t=1}^{T}\left(\frac{\widehat{\eta}_t-\widehat{\mu}}{\widehat{\sigma}}\right)^4\frac{1}{T}\\ &= \frac{1}{\widehat{\sigma}^4}\frac{1}{T}\sum_{t=1}^{T}(\widehat{\eta}_t-\widehat{\mu})^4\\ &= \kappa+o_p(1),\end{aligned}$$

where $\mathscr{E}_*$ denotes the conditional expectation given the observations X_1, ..., X_T.

Analogously to the previous section, we observe

$$\begin{aligned}\widehat{\mathbf{a}^*} &= (\mathbf{X}'\mathbf{X})^{-1}\mathbf{X}'(\mathbf{X}\widehat{\mathbf{a}}+\mathbf{u}^*)\\ &= \widehat{\mathbf{a}}+(\mathbf{X}'\mathbf{X})^{-1}\mathbf{X}'\mathbf{u}^*,\end{aligned}$$

that is,

$$\sqrt{T-p-q}(\widehat{\mathbf{a}^*}-\widehat{\mathbf{a}}) = \left(\frac{1}{T-p-q}\mathbf{X}'\mathbf{X}\right)^{-1}\left(\frac{1}{\sqrt{T-p-q}}\mathbf{X}'\mathbf{u}^*\right).$$

Analogously, let $m_t^* = \varepsilon_t^{*2}-\widetilde{h}_t$, then

$$\begin{array}{ccccc}\mathbf{e}^* & = & \widehat{\mathbf{E}} & \widetilde{\mathbf{b}} & + \quad \mathbf{m}^*,\\ \begin{pmatrix}\varepsilon_{p+q+1}^{*2}\\ \varepsilon_{p+q+2}^{*2}\\ \vdots\\ \varepsilon_T^{*2}\end{pmatrix} & = & \begin{pmatrix}1 & \widehat{\varepsilon_{p+q}}^2 & \widehat{\varepsilon_{p+q-1}}^2 & \cdots & \widehat{\varepsilon_{p+1}}^2\\ 1 & \widehat{\varepsilon_{p+q+1}}^2 & \widehat{\varepsilon_{p+q}}^2 & \cdots & \widehat{\varepsilon_{p+2}}^2\\ \vdots & \vdots & \vdots & \ddots & \vdots\\ 1 & \widehat{\varepsilon_{T-1}}^2 & \widehat{\varepsilon_{T-2}}^2 & \cdots & \widehat{\varepsilon_{T-q}}^2\end{pmatrix} & \begin{pmatrix}\widetilde{b_0}\\ \widetilde{b_1}\\ \widetilde{b_2}\\ \vdots\\ \widetilde{b_q}\end{pmatrix} & + \begin{pmatrix}m_{p+q+1}^*\\ m_{p+q+2}^*\\ \vdots\\ m_T^*\end{pmatrix}\end{array}$$

and

$$\begin{aligned}\widetilde{\mathbf{b}^*} &= (\widehat{\mathbf{E}}'\widehat{\mathbf{E}})^{-1}\widehat{\mathbf{E}}'\mathbf{e}^*\\ &= (\widehat{\mathbf{E}}'\widehat{\mathbf{E}})^{-1}\widehat{\mathbf{E}}'(\widehat{\mathbf{E}}\widetilde{\mathbf{b}}+\mathbf{m}^*)\\ &= \widetilde{\mathbf{b}}+(\widehat{\mathbf{E}}'\widehat{\mathbf{E}})^{-1}\widehat{\mathbf{E}}'\mathbf{m}^*,\end{aligned}$$

that is,

$$\sqrt{T-p-q}(\widetilde{\mathbf{b}^*}-\widetilde{\mathbf{b}}) = \left(\frac{1}{T-p-q}\widehat{\mathbf{E}}'\widehat{\mathbf{E}}\right)^{-1}\left(\frac{1}{\sqrt{T-p-q}}\widehat{\mathbf{E}}'\mathbf{m}^*\right).$$

Remark 3.5

For simplicity, we denote

$$\begin{aligned}\widetilde{h_t}-h_t =& \widehat{\mathbf{e_t}}'\widetilde{\mathbf{b}}-\mathbf{e_t}'\mathbf{b}\\ =&\mathbf{e_t}'\left(\widetilde{\mathbf{b}}-\mathbf{b}\right)+\left(\widehat{\mathbf{e_t}}'-\mathbf{e_t}'\right)\widetilde{\mathbf{b}}\\ =&\mathbf{e_t}'\left(\widetilde{\mathbf{b}}-\mathbf{b}\right)+\sum_{k=0}^{q}\left(\widehat{e_{t,k}}-e_{t,k}\right)\widetilde{b_k}\\ =&\mathbf{e_t}'\left(\widetilde{\mathbf{b}}-\mathbf{b}\right)+\sum_{k=1}^{q}\left(\widehat{\varepsilon_{t-k}}^2-\varepsilon_{t-k}^2\right)\widetilde{b_k},\end{aligned}$$

where

$$\begin{aligned}\mathbf{e_t} &= (1\ \varepsilon_{t-1}^2\ \varepsilon_{t-2}^2\ \ldots\ \varepsilon_{t-q}^2)' = (e_{t,0}\ e_{t,1}\ e_{t,2}\ \ldots\ e_{t,q})',\\ \widehat{\mathbf{e_t}} &= (1\ \widehat{\varepsilon_{t-1}}^2\ \widehat{\varepsilon_{t-2}}^2\ \ldots\ \widehat{\varepsilon_{t-q}}^2)' = (\widehat{e_{t,0}}\ \widehat{e_{t,1}}\ \widehat{e_{t,2}}\ \ldots\ \widehat{e_{t,q}})',\\ \mathbf{b} &= (b_0\ b_1\ \ldots\ b_q)' \quad and \quad \widetilde{\mathbf{b}} = (\widetilde{b_0}\ \widetilde{b_1}\ \ldots\ \widetilde{b_q})'\end{aligned}$$

throughout this chapter without further notice.

Lemma 3.1

Suppose that X_t is generated by the model (3.1) satisfying assumptions (A1)-(A4) and (B1)-(B4). Then

$$(i)\quad \frac{1}{T}\sum_{t=1}^{T}\left(\widetilde{h_t}-h_t\right)\mathbf{x_t}\mathbf{x_t}' = o_p(1),$$

$$(ii)\quad \frac{1}{T}\sum_{t=1}^{T}\left(\widetilde{h_t}^2\widehat{\mathbf{e_t}}\widehat{\mathbf{e_t}}'-h_t^2\mathbf{e_t}\mathbf{e_t}'\right) = o_p(1).$$

Proof. To prove (i), it suffices to show the following properties for every $i, j \in \mathbb{Z}$:

$$\frac{1}{T}\sum_{t=1}^{T}\left(\widetilde{h_t}-h_t\right) = o_p(1), \tag{3.16}$$

$$\frac{1}{T}\sum_{t=1}^{T}X_{t-i}\left(\widetilde{h_t}-h_t\right) = o_p(1), \tag{3.17}$$

$$\frac{1}{T}\sum_{t=1}^{T} X_{t-i}X_{t-j}\left(\widetilde{h}_t - h_t\right) = o_p(1). \tag{3.18}$$

Firstly, the left hand side of (3.16) equals

$$\begin{aligned}
&\frac{1}{T}\sum_{t=1}^{T}\left(\widetilde{h}_t - h_t\right)\\
=&\frac{1}{T}\sum_{t=1}^{T}\mathbf{e_t}'\left(\widetilde{\mathbf{b}}-\mathbf{b}\right)+\frac{1}{T}\sum_{t=1}^{T}\sum_{k=1}^{q}\left(\widehat{\varepsilon_{t-k}}^2-\varepsilon_{t-k}^2\right)\widetilde{b_k}\\
=&\left(\frac{1}{T}\sum_{t=1}^{T}\underbrace{\mathbf{e_t}'}_{=O_p(1)}\right)\underbrace{\left(\widetilde{\mathbf{b}}-\mathbf{b}\right)}_{=O_p(T^{-1/2})}+\sum_{k=1}^{q}\left(\underbrace{b_k}_{=O(1)}+\underbrace{(\widetilde{b_k}-b_k)}_{=O_p(T^{-1/2})}\right)\underbrace{\frac{1}{T}\sum_{t=1}^{T}\left(\widehat{\varepsilon_{t-k}^2}-\varepsilon_{t-k}^2\right)}_{\overset{(3.12)}{=}o_p(T^{-1/2})}\\
=&o_p(1).
\end{aligned}$$

Secondly, the left hand side of (3.17) equals

$$\begin{aligned}
&\frac{1}{T}\sum_{t=1}^{T}X_{t-i}\left(\widetilde{h}_t - h_t\right)\\
=&\left(\frac{1}{T}\sum_{t=1}^{T}\underbrace{X_{t-i}\mathbf{e_t}'}_{=O_p(1)}\right)\underbrace{\left(\widetilde{\mathbf{b}}-\mathbf{b}\right)}_{=O_p(T^{-1/2})}+\sum_{k=1}^{q}\underbrace{\widetilde{b_k}}_{=O_p(1)}\underbrace{\frac{1}{T}\sum_{t=1}^{T}X_{t-i}\left(\widehat{\varepsilon_{t-k}^2}-\varepsilon_{t-k}^2\right)}_{=o_p(1)}\\
=&o_p(1).
\end{aligned}$$

Here we obtain

$$\frac{1}{T}\sum_{t=1}^{T}X_{t-i}\left(\widehat{\varepsilon_{t-k}}^2-\varepsilon_{t-k}^2\right)=o_p(1)$$

because the left hand side equals

$$
\begin{aligned}
&\frac{1}{T}\sum_{t=1}^{T} X_{t-i}\left(\widehat{\varepsilon_{t-k}}-\varepsilon_{t-k}\right)^2+\frac{2}{T}\sum_{t=1}^{T} X_{t-i}\varepsilon_{t-k}\left(\widehat{\varepsilon_{t-k}}-\varepsilon_{t-k}\right)\\
=&\frac{1}{T}\sum_{t=1}^{T} X_{t-i}\left(\mathbf{x_{t-k}}'\left(\mathbf{a}-\widehat{\mathbf{a}}\right)\right)^2+\left(\frac{2}{T}\sum_{t=1}^{T}\underbrace{X_{t-i}\varepsilon_{t-k}\mathbf{x_{t-k}}'}_{=O_p(1)}\right)\underbrace{\left(\mathbf{a}-\widehat{\mathbf{a}}\right)}_{=O_p(T^{-1/2})}\\
\leq&\frac{1}{T}\sum_{t=1}^{T} X_{t-i}\left(\sum_{l=0}^{p} x_{t-k,l}^2\sum_{l=0}^{p}\left(a_l-\widehat{a_l}\right)^2\right)+o_p(1)\\
=&\underbrace{\sum_{l=0}^{p}\left(a_l-\widehat{a_l}\right)^2}_{=O_p(T^{-1})}\left(\frac{1}{T}\sum_{t=1}^{T}\underbrace{X_{t-i}\sum_{l=0}^{p} x_{t-k,l}^2}_{=O_p(1)}\right)+o_p(1)\\
=&o_p(1).
\end{aligned}
$$

Thirdly, the left hand side of (3.18) equals

$$
\begin{aligned}
&\frac{1}{T}\sum_{t=1}^{T} X_{t-i}X_{t-j}\left(\widetilde{h_t}-h_t\right)\\
=&\left(\frac{1}{T}\sum_{t=1}^{T}\underbrace{X_{t-i}X_{t-j}\mathbf{e_t}'}_{=O_p(1)}\underbrace{\left(\widetilde{\mathbf{b}}-\mathbf{b}\right)}_{=O_p(T^{-1/2})}\right)+\sum_{k=1}^{q}\underbrace{\widetilde{b_k}}_{=O_p(1)}\underbrace{\frac{1}{T}\sum_{t=1}^{T} X_{t-i}X_{t-j}\left(\widehat{\varepsilon_{t-k}^2}-\varepsilon_{t-k}^2\right)}_{=o_p(1)}\\
=&o_p(1).
\end{aligned}
$$

Here we obtain

$$
\frac{1}{T}\sum_{t=1}^{T} X_{t-i}X_{t-j}\left(\widehat{\varepsilon_{t-k}}^2-\varepsilon_{t-k}^2\right)=o_p(1)
$$

because the left hand side equals

$$
\begin{aligned}
&\frac{1}{T}\sum_{t=1}^{T} X_{t-i}X_{t-j}\left(\mathbf{x_{t-k}}'(\mathbf{a}-\widehat{\mathbf{a}})\right)^2 + \left(\frac{2}{T}\sum_{t=1}^{T}\underbrace{X_{t-i}X_{t-j}\varepsilon_{t-k}\mathbf{x_{t-k}}'}_{=O_p(1)}\right)\underbrace{(\mathbf{a}-\widehat{\mathbf{a}})}_{=O_p(T^{-1/2})}\\
\leq&\frac{1}{T}\sum_{t=1}^{T} X_{t-i}X_{t-j}\left(\sum_{l=0}^{p} x_{t-k,l}^2 \sum_{l=0}^{p}(a_l-\widehat{a}_l)^2\right)+o_p(1)\\
=&\underbrace{\sum_{l=0}^{p}(a_l-\widehat{a}_l)^2}_{=O_p(T^{-1})}\left(\frac{1}{T}\sum_{t=1}^{T}\underbrace{X_{t-i}X_{t-j}\sum_{l=0}^{p}x_{t-k,l}^2}_{=O_p(1)}\right)+o_p(1)\\
=&o_p(1).
\end{aligned}
$$

To prove (ii), it suffices to show the following properties for every $i, j \in \mathbb{Z}$:

$$\frac{1}{T}\sum_{t=1}^{T}\left(\widetilde{h}_t^2 - h_t^2\right) = o_p(1), \tag{3.19}$$

$$\frac{1}{T}\sum_{t=1}^{T}\left(\widetilde{h}_t^2\widehat{\varepsilon_{t-i}}^2 - h_t^2\varepsilon_{t-i}^2\right) = o_p(1), \tag{3.20}$$

$$\frac{1}{T}\sum_{t=1}^{T}\left(\widetilde{h}_t^2\widehat{\varepsilon_{t-i}}^2\widehat{\varepsilon_{t-j}}^2 - h_t^2\varepsilon_{t-i}^2\varepsilon_{t-j}^2\right) = o_p(1). \tag{3.21}$$

Firstly, the left hand side of (3.19) equals

$$\frac{1}{T}\sum_{t=1}^{T}\left(\widetilde{h}_t^2 - h_t^2\right) = \frac{1}{T}\sum_{t=1}^{T}\left(\widetilde{h}_t - h_t\right)^2 + \frac{2}{T}\sum_{t=1}^{T}h_t\left(\widetilde{h}_t - h_t\right) = o_p(1)$$

because

$$
\begin{aligned}
&\frac{2}{T}\sum_{t=1}^{T}h_t\left(\widetilde{h}_t - h_t\right)\\
=&\left(\frac{2}{T}\sum_{t=1}^{T}\underbrace{h_t\mathbf{e_t}'}_{=O_p(1)}\right)\underbrace{\left(\widetilde{\mathbf{b}}-\mathbf{b}\right)}_{=O_p(T^{-1/2})} + \sum_{k=1}^{q}\underbrace{\widetilde{b}_k}_{=O_p(1)}\underbrace{\frac{1}{T}\sum_{t=1}^{T}h_t\left(\widehat{\varepsilon_{t-k}}^2 - \varepsilon_{t-k}^2\right)}_{=o_p(1)}\\
=&o_p(1)
\end{aligned}
$$

and

$$\begin{aligned}
&\frac{1}{T}\sum_{t=1}^{T}\left(\widetilde{h}_t - h_t\right)^2 \\
=&\frac{1}{T}\sum_{t=1}^{T}\left(\left(\mathbf{e_t}'(\widetilde{\mathbf{b}}-\mathbf{b})\right)^2 + 2\sum_{k=1}^{q}\mathbf{e_t}'\left(\widetilde{\mathbf{b}}-\mathbf{b}\right)\left(\widehat{\varepsilon_{t-k}}^2 - \varepsilon_{t-k}^2\right)\widetilde{b_k}\right. \\
&\left.+\left(\sum_{k=1}^{q}(\widehat{\varepsilon_{t-k}}^2 - \varepsilon_{t-k}^2)\widetilde{b_k}\right)^2\right) \\
\leq&\underbrace{\left(\frac{1}{T}\sum_{t=1}^{T}\sum_{k=0}^{q}e_{t,\,k}^2\right)}_{=O_p(1)}\underbrace{\sum_{k=0}^{q}(\widetilde{b_k}-b_k)^2}_{=O_p(T^{-1})} + 2\left(\widetilde{\mathbf{b}}-\mathbf{b}\right)\frac{1}{T}\sum_{t=1}^{T}\mathbf{e_t}'\left(\sum_{k=1}^{q}\widetilde{b_k}(\widehat{\varepsilon_{t-k}}^2 - \varepsilon_{t-k}^2)\right) \\
&+\underbrace{\left(\frac{1}{T}\sum_{t=1}^{T}\sum_{k=1}^{q}\left(\widehat{\varepsilon_{t-k}}^2 - \varepsilon_{t-k}^2\right)^2\right)}_{\overset{(3.13),\,(3.14)}{=}o_p(1)}\underbrace{\sum_{k=1}^{q}\widetilde{b_k}^2}_{=O_p(1)} \\
=&o_p(1) + 2\underbrace{\left(\widetilde{\mathbf{b}}-\mathbf{b}\right)}_{=O_p(T^{-1/2})}\sum_{k=1}^{q}\underbrace{\widetilde{b_k}}_{=O_p(1)}\underbrace{\frac{1}{T}\sum_{t=1}^{T}\mathbf{e_t}'\left(\widehat{\varepsilon_{t-k}}^2 - \varepsilon_{t-k}^2\right)}_{\overset{(3.12),\,(3.13)}{=}o_p(T^{-1/2})} \\
=&o_p(1).
\end{aligned}$$

Secondly, to prove (3.20) it suffices to show

$$\frac{1}{T}\sum_{t=1}^{T}\varepsilon_{t-i}^2\left(\widetilde{h_t}^2 - h_t^2\right) = o_p(1), \tag{3.22}$$

$$\frac{1}{T}\sum_{t=1}^{T}\left(\widetilde{h_t}^2 - h_t^2\right)\left(\widehat{\varepsilon_{t-i}}^2 - \varepsilon_{t-i}^2\right) = o_p(1), \tag{3.23}$$

$$\frac{1}{T}\sum_{t=1}^{T}h_t^2\left(\widehat{\varepsilon_{t-i}}^2 - \varepsilon_{t-i}^2\right) = o_p(1). \tag{3.24}$$

The left hand side of (3.22) equals

$$\frac{1}{T}\sum_{t=1}^{T}\varepsilon_{t-i}^2\left(\widetilde{h_t}^2 - h_t^2\right) = \frac{1}{T}\sum_{t=1}^{T}\varepsilon_{t-i}^2\left(\widetilde{h}_t - h_t\right)^2 + \frac{2}{T}\sum_{t=1}^{T}\varepsilon_{t-i}^2 h_t\left(\widetilde{h}_t - h_t\right) = o_p(1)$$

because

$$
\begin{aligned}
&\frac{2}{T}\sum_{t=1}^{T}\varepsilon_{t-i}^{2}h_t\left(\widetilde{h_t}-h_t\right)\\
=&\left(\frac{2}{T}\sum_{t=1}^{T}\underbrace{\varepsilon_{t-i}^{2}h_t\mathbf{e_t}'}_{=O_p(1)}\right)\underbrace{\left(\widetilde{\mathbf{b}}-\mathbf{b}\right)}_{=O_p(T^{-1/2})}+\sum_{k=1}^{q}\underbrace{\widetilde{b_k}}_{=O_p(1)}\underbrace{\frac{1}{T}\sum_{t=1}^{T}\varepsilon_{t-i}^{2}h_t\left(\widehat{\varepsilon_{t-k}}^{2}-\varepsilon_{t-k}^{2}\right)}_{=o_p(1)}\\
=&o_p(1)
\end{aligned}
$$

and

$$
\begin{aligned}
&\frac{1}{T}\sum_{t=1}^{T}\varepsilon_{t-i}^{2}\left(\widetilde{h_t}-h_t\right)^2\\
=&\frac{1}{T}\sum_{t=1}^{T}\varepsilon_{t-i}^{2}\left(\left(\mathbf{e_t}'(\widetilde{\mathbf{b}}-\mathbf{b})\right)^2+2\sum_{k=1}^{q}\mathbf{e_t}'\left(\widetilde{\mathbf{b}}-\mathbf{b}\right)\left(\widehat{\varepsilon_{t-k}}^{2}-\varepsilon_{t-k}^{2}\right)\widetilde{b_k}\right.\\
&\left.+\left(\sum_{k=1}^{q}(\widehat{\varepsilon_{t-k}}^{2}-\varepsilon_{t-k}^{2})\widetilde{b_k}\right)^2\right)\\
\leq&\underbrace{\left(\frac{1}{T}\sum_{t=1}^{T}\varepsilon_{t-i}^{2}\sum_{k=0}^{q}e_{t,\,k}^{2}\right)}_{=O_p(1)}\underbrace{\sum_{k=0}^{q}\left(\widetilde{b_k}-b_k\right)^2}_{=O_p(T^{-1})}\\
&+2\left(\widetilde{\mathbf{b}}-\mathbf{b}\right)\frac{1}{T}\sum_{t=1}^{T}\varepsilon_{t-i}^{2}\mathbf{e_t}'\left(\sum_{k=1}^{q}\widetilde{b_k}(\widehat{\varepsilon_{t-k}}^{2}-\varepsilon_{t-k}^{2})\right)\\
&+\underbrace{\left(\frac{1}{T}\sum_{t=1}^{T}\varepsilon_{t-i}^{2}\sum_{k=1}^{q}\left(\widehat{\varepsilon_{t-k}}^{2}-\varepsilon_{t-k}^{2}\right)^2\right)}_{=o_p(1)}\underbrace{\sum_{k=1}^{q}\widetilde{b_k}^{2}}_{=O_p(1)}\\
=&o_p(1)+2\underbrace{\left(\widetilde{\mathbf{b}}-\mathbf{b}\right)}_{=O_p(T^{-1/2})}\sum_{k=1}^{q}\underbrace{\widetilde{b_k}}_{=O_p(1)}\underbrace{\frac{1}{T}\sum_{t=1}^{T}\varepsilon_{t-i}^{2}\mathbf{e_t}'\left(\widehat{\varepsilon_{t-k}}^{2}-\varepsilon_{t-k}^{2}\right)}_{=o_p(1)}\\
=&o_p(1).
\end{aligned}
$$

The left hand side of (3.23) equals

$$\begin{aligned}
&\frac{1}{T}\sum_{t=1}^{T}\left(\widehat{\varepsilon_{t-i}}^2-\varepsilon_{t-i}^2\right)\left(\widetilde{h_t}^2-h_t^2\right)\\
=&\frac{1}{T}\sum_{t=1}^{T}\left(\widehat{\varepsilon_{t-i}}^2-\varepsilon_{t-i}^2\right)\left(\widetilde{h_t}-h_t\right)^2+\frac{2}{T}\sum_{t=1}^{T}\left(\widehat{\varepsilon_{t-i}}^2-\varepsilon_{t-i}^2\right)h_t\left(\widetilde{h_t}-h_t\right)\\
=&o_p(1)
\end{aligned}$$

because

$$\begin{aligned}
&\frac{1}{T}\sum_{t=1}^{T}\left(\widehat{\varepsilon_{t-i}}^2-\varepsilon_{t-i}^2\right)\left(\widetilde{h_t}-h_t\right)^2\\
=&\frac{1}{T}\sum_{t=1}^{T}\left(\widehat{\varepsilon_{t-i}}^2-\varepsilon_{t-i}^2\right)\left(\left(\mathbf{e_t}'(\widetilde{\mathbf{b}}-\mathbf{b})\right)^2+2\sum_{k=1}^{q}\mathbf{e_t}'\left(\widetilde{\mathbf{b}}-\mathbf{b}\right)\left(\widehat{\varepsilon_{t-k}}^2-\varepsilon_{t-k}^2\right)\widetilde{b_k}\right.\\
&\left.\quad+\left(\sum_{k=1}^{q}(\widehat{\varepsilon_{t-k}}^2-\varepsilon_{t-k}^2)\widetilde{b_k}\right)^2\right)\\
\leq&\underbrace{\left(\frac{1}{T}\sum_{t=1}^{T}\left(\widehat{\varepsilon_{t-i}}^2-\varepsilon_{t-i}^2\right)\sum_{k=0}^{q}e_{t,\,k}^2\right)}_{=O_p(1)}\underbrace{\sum_{k=0}^{q}(\widetilde{b_k}-b_k)^2}_{=O_p(T^{-1})}\\
&+2\left(\widetilde{\mathbf{b}}-\mathbf{b}\right)\frac{1}{T}\sum_{t=1}^{T}\left(\widehat{\varepsilon_{t-i}}^2-\varepsilon_{t-i}^2\right)\mathbf{e_t}'\left(\sum_{k=1}^{q}\widetilde{b_k}(\widehat{\varepsilon_{t-k}}^2-\varepsilon_{t-k}^2)\right)\\
&+\underbrace{\left(\frac{1}{T}\sum_{t=1}^{T}\left(\widehat{\varepsilon_{t-i}}^2-\varepsilon_{t-i}^2\right)\sum_{k=1}^{q}\left(\widehat{\varepsilon_{t-k}}^2-\varepsilon_{t-k}^2\right)^2\right)}_{=o_p(1)}\underbrace{\sum_{k=1}^{q}\widetilde{b_k}^2}_{=O_p(1)}\\
=&o_p(1)+2\underbrace{\left(\widetilde{\mathbf{b}}-\mathbf{b}\right)}_{=O_p(T^{-1/2})}\sum_{k=1}^{q}\underbrace{\widetilde{b_k}}_{=O_p(1)}\underbrace{\frac{1}{T}\sum_{t=1}^{T}\left(\widehat{\varepsilon_{t-i}}^2-\varepsilon_{t-i}^2\right)\mathbf{e_t}'\left(\widehat{\varepsilon_{t-k}}^2-\varepsilon_{t-k}^2\right)}_{=o_p(1)}\\
=&o_p(1)
\end{aligned}$$

and

$$
\begin{aligned}
&\frac{2}{T}\sum_{t=1}^{T}\left(\widehat{\varepsilon_{t-i}}^2-\varepsilon_{t-i}^2\right)h_t\left(\widetilde{h_t}-h_t\right)\\
=&\underbrace{\left(\frac{2}{T}\sum_{t=1}^{T}\left(\widehat{\varepsilon_{t-i}}^2-\varepsilon_{t-i}^2\right)h_t\mathbf{e_t}'\right)}_{=O_p(1)}\underbrace{\left(\widetilde{\mathbf{b}}-\mathbf{b}\right)}_{=O_p(T^{-1/2})}\\
&+\sum_{k=1}^{q}\underbrace{\widetilde{b_k}}_{=O_p(1)}\underbrace{\frac{1}{T}\sum_{t=1}^{T}\left(\widehat{\varepsilon_{t-i}}^2-\varepsilon_{t-i}^2\right)h_t\left(\widehat{\varepsilon_{t-k}}^2-\varepsilon_{t-k}^2\right)}_{=o_p(1)}\\
=&o_p(1).
\end{aligned}
$$

The left hand side of (3.24) equals

$$
\begin{aligned}
&\frac{1}{T}\sum_{t=1}^{T}h_t^2\left(\widehat{\varepsilon_{t-i}}-\varepsilon_{t-i}\right)^2+\frac{2}{T}\sum_{t=1}^{T}h_t^2\varepsilon_{t-i}\left(\widehat{\varepsilon_{t-i}}-\varepsilon_{t-i}\right)\\
=&\frac{1}{T}\sum_{t=1}^{T}h_t^2\left(\mathbf{x_{t-i}}'(\mathbf{a}-\widehat{\mathbf{a}})\right)^2+\left(\frac{2}{T}\sum_{t=1}^{T}\underbrace{h_t^2\varepsilon_{t-i}\mathbf{x_{t-i}}'}_{=O_p(1)}\right)\underbrace{(\mathbf{a}-\widehat{\mathbf{a}})}_{=O_p(T^{-1/2})}\\
\leq&\frac{1}{T}\sum_{t=1}^{T}h_t^2\left(\sum_{k=0}^{p}x_{t-i,\,k}^2\sum_{k=0}^{p}(a_k-\widehat{a_k})^2\right)+o_p(1)\\
=&\underbrace{\sum_{k=0}^{p}(a_k-\widehat{a_k})^2}_{=O_p(T^{-1})}\left(\frac{1}{T}\sum_{t=1}^{T}\underbrace{h_t^2\sum_{k=0}^{p}x_{t-i,\,k}^2}_{=O_p(1)}\right)+o_p(1)\\
=&o_p(1).
\end{aligned}
$$

The equation (3.21) can be proved analogously. □

Lemma 3.2

Suppose that X_t is generated by the model (3.1) satisfying assumptions (A1)-(A4) and (B1)-(B4). Let $\mathbf{c_1}' = (c_{1,0}, c_{1,1}, c_{1,2}, ..., c_{1,p}) \in \mathbb{R}^{p+1}$ and $\mathbf{c_2}' = (c_{2,0}, c_{2,1}, c_{2,2}, ..., c_{2,q}) \in \mathbb{R}^{q+1}$, then

$$(i) \quad \frac{1}{T-p-q}\sum_{t=p+q+1}^{T}\left|\mathbf{c_1}'\sqrt{\widetilde{h}_t}\mathbf{x_t}\right|^3 = O_p(1),$$

$$(ii) \quad \frac{1}{T-p-q}\sum_{t=p+q+1}^{T}\left|\mathbf{c_2}'\widetilde{h}_t\widehat{\mathbf{e}}_{\mathbf{t}}\right|^3 = O_p(1).$$

Proof. Since $|a+b| \leq |a|+|b|$, we have

$$|a+b|^3 \leq 8\max\{|a|^3,|b|^3\} \leq 8\big(|a|^3+|b|^3\big),$$

therefore,

$$\begin{aligned}
&\frac{1}{T-p-q}\sum_{t=p+q+1}^{T}\left|\mathbf{c_1}'\sqrt{\widetilde{h}_t}\mathbf{x_t}\right|^3 \\
\leq&\frac{8}{T-p-q}\sum_{t=p+q+1}^{T}\left(\left|\mathbf{c_1}'(\sqrt{\widetilde{h}_t}-\sqrt{h_t})\mathbf{x_t}\right|^3 + \underbrace{\left|\mathbf{c_1}'\sqrt{h_t}\mathbf{x_t}\right|^3}_{\overset{(A3)}{=}O_p(1)}\right) \\
=&\frac{8}{T-p-q}\sum_{t=p+q+1}^{T}\left|\sum_{k=0}^{p}c_{1,k}(\sqrt{\widetilde{h}_t}-\sqrt{h_t})x_{t,k}\right|^3 + O_p(1) \\
\leq&\frac{8}{T-p-q}\sum_{t=p+q+1}^{T}8^p\sum_{k=0}^{p}|c_{1,k}|^3\left|(\sqrt{\widetilde{h}_t}-\sqrt{h_t})x_{t,k}\right|^3 + O_p(1) \\
\leq&\Big(\min\{b_0,\widetilde{b_0}\}\Big)^{-3/2}8^p\sum_{k=0}^{p}|c_{1,k}|^3\left(\frac{1}{T-p-q}\sum_{t=p+q+1}^{T}\underbrace{\left|(\widetilde{h}_t-h_t)x_{t,k}\right|^3}_{=O_p(1)}\right)+O_p(1) \\
=&O_p(1).
\end{aligned}$$

Analogously,

$$
\begin{aligned}
&\frac{1}{T-p-q}\sum_{t=p+q+1}^{T}\left|\mathbf{c_2}'\widetilde{h}_t\widehat{\mathbf{e}}_\mathbf{t}\right|^3 \\
\leq&\frac{64}{T-p-q}\sum_{t=p+q+1}^{T}\Bigg(\left|\mathbf{c_2}'(\widetilde{h}_t-h_t)\mathbf{e_t}\right|^3+\left|\mathbf{c_2}'(\widetilde{h}_t-h_t)(\widehat{\mathbf{e}}_\mathbf{t}-\mathbf{e_t})\right|^3 \\
&+\underbrace{\left|\mathbf{c_2}'h_t\mathbf{e_t}\right|^3}_{\overset{(B3)}{=}O_p(1)}+\left|\mathbf{c_2}'h_t(\widehat{\mathbf{e}}_\mathbf{t}-\mathbf{e_t})\right|^3\Bigg) \\
=&O_p(1).
\end{aligned}
$$

Here we obtain the last equation because

$$
\begin{aligned}
&\frac{64}{T-p-q}\sum_{t=p+q+1}^{T}8^q\sum_{k=0}^{q}|c_{2,k}|^3\underbrace{\left|(\widetilde{h}_t-h_t)e_{t,k}\right|^3}_{=O_p(1)} \\
+&\frac{64}{T-p-q}\sum_{t=p+q+1}^{T}8^{q-1}\sum_{k=1}^{q}|c_{2,k}|^3\underbrace{\left|(\widetilde{h}_t-h_t)(\widehat{\varepsilon_{t-k}}^2-\varepsilon_{t-k}^2)\right|^3}_{=O_p(1)} \\
+&\frac{64}{T-p-q}\sum_{t=p+q+1}^{T}8^{q-1}\sum_{k=1}^{q}|c_{2,k}|^3\underbrace{\left|h_t(\widehat{\varepsilon_{t-k}}^2-\varepsilon_{t-k}^2)\right|^3}_{=O_p(1)}+O_p(1) \\
=&O_p(1).
\end{aligned}
$$

□

Theorem 3.3

Suppose that X_t is generated by the model (3.1) satisfying assumptions (A1)-(A4) and (B1)-(B4). Then the residual bootstrap of section 3.2 for the AR part is weakly consistent, i.e.

$$\sqrt{T-p-q}(\widehat{\mathbf{a}}^*-\widehat{\mathbf{a}})\xrightarrow{d}\mathcal{N}\left(0,\mathbf{A}^{-1}\mathbf{V}\mathbf{A}^{-1}\right)\quad \textit{in probability.}$$

Proof. From Theorem 3.1 we already have

$$\frac{1}{T-p-q}\mathbf{X}'\mathbf{X}\xrightarrow{p}\mathbf{A}.$$

Together with the Slutsky theorem it suffices to show

$$\frac{1}{\sqrt{T-p-q}}\mathbf{X}'\mathbf{u}^* = \frac{1}{\sqrt{T-p-q}}\sum_{t=p+q+1}^{T} \varepsilon_t^* \mathbf{x_t} \xrightarrow{d} \mathcal{N}\left(0,\mathbf{V}\right) \quad \textit{in probability.}$$

Now we observe

$$\mathbf{y_t^*} := \frac{1}{\sqrt{T-p-q}}\varepsilon_t^* \mathbf{x_t} = \frac{1}{\sqrt{T-p-q}}\sqrt{\widetilde{h}_t}\eta_t^* \mathbf{x_t}, \quad t = p+q+1, \ ..., \ T.$$

For each T the sequence $\mathbf{y_1^*},...,\mathbf{y_T^*}$ is independent because $\eta_1^*,...,\eta_T^*$ is independent. Here we obtain

$$\mathcal{E}_*(\mathbf{y_t^*}) = \frac{1}{\sqrt{T-p-q}}\sqrt{\widetilde{h}_t}\mathbf{x_t}\mathcal{E}_*(\eta_t^*) = 0$$

and

$$\begin{aligned}
&\sum_{t=p+q+1}^{T} \mathcal{E}_*\left(\mathbf{y_t^*}\mathbf{y_t^*}'\right) \\
=&\frac{1}{T-p-q}\sum_{t=p+q+1}^{T} \widetilde{h}_t \mathbf{x_t}\mathbf{x_t}'\mathcal{E}_*(\eta_t^{*2}) \\
=&\frac{1}{T-p-q}\sum_{t=p+q+1}^{T} h_t \mathbf{x_t}\mathbf{x_t}' + \underbrace{\frac{1}{T-p-q}\sum_{t=p+q+1}^{T}\left(\widetilde{h}_t - h_t\right)\mathbf{x_t}\mathbf{x_t}'}_{\overset{Lemma\ 3.1}{=} o_p(1)} \\
&\xrightarrow{p}\mathbf{V}.
\end{aligned}$$

Let $\mathbf{c} \in \mathbb{R}^{p+q+1}$ and

$$s_T^2 := \sum_{t=p+q+1}^{T} \mathcal{E}_*\left(\mathbf{c}'\mathbf{y_t^*}\right)^2 \xrightarrow{p} \mathbf{c}'\mathbf{V}\mathbf{c},$$

then we have the Lyapounov condition for $\delta = 1$

$$\begin{aligned}
&\sum_{t=p+q+1}^{T} \frac{1}{s_T^3}\mathcal{E}_*\left|\mathbf{c}'\mathbf{y_t^*}\right|^3 \\
=&\frac{\mathcal{E}_*|\eta_t^*|^3}{s_T^3(T-p-q)^{1/2}}\underbrace{\frac{1}{T-p-q}\sum_{t=p+q+1}^{T}\left|\mathbf{c}'\sqrt{\widetilde{h}_t}\mathbf{x_t}\right|^3}_{\overset{Lemma\ 3.2}{=} O_p(1)} \\
=&o_p(1).
\end{aligned}$$

The Cramér-Wold theorem and the central limit theorem for triangular arrays applied to the sequence $\mathbf{c'y_1^*}, ..., \mathbf{c'y_T^*}$ show

$$\sum_{t=p+q+1}^{T} \mathbf{y_t^*} \xrightarrow{d} \mathscr{N}\left(0, \mathbf{V}\right) \quad \textit{in probability.}$$

□

Theorem 3.4

Suppose that X_t is generated by the model (3.1) satisfying assumptions (A1)-(A4) and (B1)-(B4). Then the residual bootstrap of section 3.2 for the ARCH part is weakly consistent, i.e.

$$\sqrt{T-p-q}(\widetilde{\mathbf{b}^*} - \widetilde{\mathbf{b}}) \xrightarrow{d} \mathscr{N}\left(0, \mathbf{B}^{-1}\mathbf{W}\mathbf{B}^{-1}\right) \quad \textit{in probability.}$$

Proof. From Theorem 3.2 we already have

$$\frac{1}{T-p-q}\widehat{\mathbf{E}}'\widehat{\mathbf{E}} \xrightarrow{p} \mathbf{B}.$$

Together with the Slutsky theorem it suffices to show

$$\frac{1}{\sqrt{T-p-q}}\widehat{\mathbf{E}}'\mathbf{m}^* = \frac{1}{\sqrt{T-p-q}} \sum_{t=p+q+1}^{T} m_t^* \widehat{\mathbf{e_t}} \xrightarrow{d} \mathscr{N}\left(0, \mathbf{W}\right) \quad \textit{in probability.}$$

Now we observe

$$\mathbf{z_t^*} := \frac{1}{\sqrt{T-p-q}} m_t^* \widehat{\mathbf{e_t}} = \frac{1}{\sqrt{T-p-q}} \widetilde{h}_t (\eta_t^{*2} - 1)\widehat{\mathbf{e_t}}, \quad t = p+q+1, \ ..., \ T.$$

For each T the sequence $\mathbf{z_1^*}, ..., \mathbf{z_T^*}$ is independent because $\eta_1^*, ..., \eta_T^*$ is independent. Here we obtain

$$\mathscr{E}_*\left(\mathbf{z_t^*}\right) = \frac{1}{\sqrt{T-p-q}} \widetilde{h}_t \widehat{\mathbf{e_t}} \mathscr{E}_*(\eta_t^{*2} - 1) = 0$$

and from Lemma 3.1

$$\begin{aligned}
&\sum_{t=p+q+1}^{T} \mathscr{E}_*\left(\mathbf{z_t^*}\mathbf{z_t^*}'\right) \\
=&\frac{1}{T-p-q} \sum_{t=p+q+1}^{T} \left(\mathscr{E}_*\left(\eta_t^{*4}\right) - 2\mathscr{E}_*(\eta_t^{*2}) + 1\right)\widetilde{h}_t^2 \widehat{\mathbf{e_t}}\widehat{\mathbf{e_t}}' \\
=&\left(\kappa - 1 + o_P(1)\right)\left(\frac{1}{T-p-q} \sum_{t=p+q+1}^{T} h_t^2 \mathbf{e_t}\mathbf{e_t}' + o_p(1)\right) \\
\xrightarrow{p}&\mathbf{W}.
\end{aligned}$$

Let $\mathbf{c} \in \mathbb{R}^{p+q+1}$ and

$$s_T^2 := \sum_{t=p+q+1}^{T} \mathscr{E}_*\left(\mathbf{c}'\mathbf{z_t^*}\right)^2 \xrightarrow{p} \mathbf{c}'\mathbf{W}\mathbf{c},$$

then we have the Lyapounov condition for $\delta = 1$

$$\begin{aligned}
&\sum_{t=p+q+1}^{T} \frac{1}{s_T^3} \mathscr{E}_* \left|\mathbf{c}'\mathbf{z_t^*}\right|^3 \\
=&\sum_{t=p+q+1}^{T} \frac{1}{s_T^3} \mathscr{E}_* \left|\mathbf{c}' \frac{1}{\sqrt{T-p-q}} \widetilde{h}_t (\eta_t^{*2} - 1)\widehat{\mathbf{e_t}}\right|^3 \\
=&\frac{\mathscr{E}_*|\eta_t^{*2} - 1|^3}{s_T^3 (T-p-q)^{1/2}} \underbrace{\frac{1}{T-p-q} \sum_{t=p+q+1}^{T} \left|\mathbf{c}'\widetilde{h}_t \widehat{\mathbf{e_t}}\right|^3}_{\overset{Lemma\ 3.2}{=} O_p(1)} \\
=&o_p(1).
\end{aligned}$$

The Cramér-Wold theorem and the central limit theorem for triangular arrays applied to the sequence $\mathbf{c}'\mathbf{z_1^*}, ..., \mathbf{c}'\mathbf{z_T^*}$ show

$$\sum_{t=p+q+1}^{T} \mathbf{z_t^*} \xrightarrow{d} \mathscr{N}\left(0, \mathbf{W}\right) \quad \textit{in probability.}$$

□

3.3 Wild Bootstrap

Analogously to the previous section, we propose a wild bootstrap technique and investigate its consistency. A wild bootstrap method can be applied to the model (3.1) as follows:

(step 1) Obtain the OLS estimator $\widehat{\mathbf{a}}$ and calculate the residuals

$$\widehat{\varepsilon}_t = X_t - \widehat{a}_0 - \sum_{i=1}^{p} \widehat{a}_i X_{t-i}, \quad t = p+1, \ ..., \ T.$$

(step 2) Compute the OLS estimator $\widetilde{\mathbf{b}}$ and calculate the estimated heteroscedasticity

$$\widetilde{h}_t = \widetilde{b}_0 + \sum_{j=1}^{q} \widetilde{b}_j \widehat{\varepsilon_{t-j}}^2, \qquad t = p+q+1, \ldots, T.$$

(step 3) Generate the bootstrap process $X_t^\dagger$ by computing

$$X_t^\dagger = \widehat{a}_0 + \sum_{i=1}^{p} \widehat{a}_i X_{t-i} + \varepsilon_t^\dagger$$

$$\varepsilon_t^\dagger = \sqrt{\widetilde{h}_t} w_t^\dagger, \quad w_t^\dagger \overset{iid}{\sim} \mathcal{N}(0,1), \quad t = p+q+1, \ldots, T,$$

or in vector and matrix notation

$$\begin{array}{ccccc} \mathbf{x}^\dagger & = & \mathbf{X} & \widehat{\mathbf{a}} & + \quad \mathbf{u}^\dagger, \end{array}$$

$$\begin{pmatrix} X_{p+q+1}^\dagger \\ X_{p+q+2}^\dagger \\ \vdots \\ X_T^\dagger \end{pmatrix} = \begin{pmatrix} 1 & X_{p+q} & X_{p+q-1} & \cdots & X_{q+1} \\ 1 & X_{p+q+1} & X_{p+q} & \cdots & X_{q+2} \\ \vdots & \vdots & \vdots & \ddots & \vdots \\ 1 & X_{T-1} & X_{T-2} & \cdots & X_{T-p} \end{pmatrix} \begin{pmatrix} \widehat{a}_0 \\ \widehat{a}_1 \\ \widehat{a}_2 \\ \vdots \\ \widehat{a}_p \end{pmatrix} + \begin{pmatrix} \varepsilon_{p+q+1}^\dagger \\ \varepsilon_{p+q+2}^\dagger \\ \vdots \\ \varepsilon_T^\dagger \end{pmatrix}.$$

(step 4) Calculate the bootstrap estimator

$$\widehat{\mathbf{a}^\dagger} = (\mathbf{X}'\mathbf{X})^{-1}\mathbf{X}'\mathbf{x}^\dagger,$$
$$\widetilde{\mathbf{b}^\dagger} = (\widehat{\mathbf{E}}'\widehat{\mathbf{E}})^{-1}\widehat{\mathbf{E}}'\mathbf{e}^\dagger,$$

where $\mathbf{e}^\dagger = \left(\varepsilon_{p+q+1}^{\dagger 2}\ \varepsilon_{p+q+2}^{\dagger 2} \cdots \varepsilon_T^{\dagger 2}\right)'$.

Analogously to the previous section, we observe

$$\widehat{\mathbf{a}^\dagger} = (\mathbf{X}'\mathbf{X})^{-1}\mathbf{X}'(\mathbf{X}\widehat{\mathbf{a}} + \mathbf{u}^\dagger)$$
$$= \widehat{\mathbf{a}} + (\mathbf{X}'\mathbf{X})^{-1}\mathbf{X}'\mathbf{u}^\dagger,$$

that is,

$$\sqrt{T-p-q}(\widehat{\mathbf{a}^\dagger} - \widehat{\mathbf{a}}) = \left(\frac{1}{T-p-q}\mathbf{X}'\mathbf{X}\right)^{-1} \left(\frac{1}{\sqrt{T-p-q}}\mathbf{X}'\mathbf{u}^\dagger\right).$$

Let $m_t^\dagger = \varepsilon_t^{\dagger 2} - \widetilde{h}_t$, then

$$\begin{array}{ccccccc} \mathbf{e}^\dagger & = & \widehat{\mathbf{E}} & \widetilde{\mathbf{b}} & + & \mathbf{m}^\dagger, \end{array}$$

$$\begin{pmatrix} \varepsilon_{p+q+1}^{\dagger 2} \\ \varepsilon_{p+q+2}^{\dagger 2} \\ \vdots \\ \varepsilon_T^{\dagger 2} \end{pmatrix} = \begin{pmatrix} 1 & \widehat{\varepsilon_{p+q}}^2 & \widehat{\varepsilon_{p+q-1}}^2 & \dots & \widehat{\varepsilon_{p+1}}^2 \\ 1 & \widehat{\varepsilon_{p+q+1}}^2 & \widehat{\varepsilon_{p+q}}^2 & \dots & \widehat{\varepsilon_{p+2}}^2 \\ \vdots & \vdots & \vdots & \ddots & \vdots \\ 1 & \widehat{\varepsilon_{T-1}}^2 & \widehat{\varepsilon_{T-2}}^2 & \dots & \widehat{\varepsilon_{T-q}}^2 \end{pmatrix} \begin{pmatrix} \widetilde{b}_0 \\ \widetilde{b}_1 \\ \widetilde{b}_2 \\ \vdots \\ \widetilde{b}_q \end{pmatrix} + \begin{pmatrix} m_{p+q+1}^\dagger \\ m_{p+q+2}^\dagger \\ \vdots \\ m_T^\dagger \end{pmatrix}$$

and

$$\begin{aligned} \widetilde{\mathbf{b}}^\dagger &= (\widehat{\mathbf{E}}'\widehat{\mathbf{E}})^{-1}\widehat{\mathbf{E}}'\mathbf{e}^\dagger \\ &= (\widehat{\mathbf{E}}'\widehat{\mathbf{E}})^{-1}\widehat{\mathbf{E}}'(\widehat{\mathbf{E}}\widetilde{\mathbf{b}} + \mathbf{m}^\dagger) \\ &= \widetilde{\mathbf{b}} + (\widehat{\mathbf{E}}'\widehat{\mathbf{E}})^{-1}\widehat{\mathbf{E}}'\mathbf{m}^\dagger, \end{aligned}$$

that is,

$$\sqrt{T-p-q}(\widetilde{\mathbf{b}}^\dagger - \widetilde{\mathbf{b}}) = \left(\frac{1}{T-p-q}\widehat{\mathbf{E}}'\widehat{\mathbf{E}}\right)^{-1}\left(\frac{1}{\sqrt{T-p-q}}\widehat{\mathbf{E}}'\mathbf{m}^\dagger\right).$$

Theorem 3.5
Suppose that X_t is generated by the model (3.1) satisfying assumptions (A1)-(A4) and (B1)-(B4). Then the wild bootstrap of section 3.3 for the AR part is weakly consistent, i.e.

$$\sqrt{T-p-q}(\widehat{\mathbf{a}}^\dagger - \widehat{\mathbf{a}}) \xrightarrow{d} \mathcal{N}\left(0, \mathbf{A}^{-1}\mathbf{V}\mathbf{A}^{-1}\right) \quad \textit{in probability.}$$

Proof. The proof is the mirror image of Theorem 3.3. From Theorem 3.1 we already have

$$\frac{1}{T-p-q}\mathbf{X}'\mathbf{X} \xrightarrow{p} \mathbf{A}.$$

Together with the Slutsky theorem it suffices to show

$$\frac{1}{\sqrt{T-p-q}}\mathbf{X}'\mathbf{u}^\dagger = \frac{1}{\sqrt{T-p-q}}\sum_{t=p+q+1}^{T} \varepsilon_t^\dagger \mathbf{x_t} \xrightarrow{d} \mathcal{N}\left(0, \mathbf{V}\right) \quad \textit{in probability.}$$

Now we observe

$$\mathbf{y_t^\dagger} := \frac{1}{\sqrt{T-p-q}}\varepsilon_t^\dagger \mathbf{x_t} = \frac{1}{\sqrt{T-p-q}}\sqrt{\widetilde{h}_t}w_t^\dagger \mathbf{x_t}, \quad t = p+q+1, \dots, T.$$

For each T the sequence $\mathbf{y_1^\dagger}, ..., \mathbf{y_T^\dagger}$ is independent because $w_1^\dagger, ..., w_T^\dagger$ is independent. Here we obtain

$$\mathscr{E}_\dagger(\mathbf{y_t^\dagger}) = \frac{1}{\sqrt{T-p-q}} \sqrt{\widetilde{h}_t} \mathbf{x_t} \mathscr{E}_\dagger(w_t^\dagger) = 0$$

and

$$\begin{aligned}
&\sum_{t=p+q+1}^{T} \mathscr{E}_\dagger\left(\mathbf{y_t^\dagger} \mathbf{y_t^\dagger}'\right) \\
=&\frac{1}{T-p-q} \sum_{t=p+q+1}^{T} \widetilde{h}_t \mathbf{x_t} \mathbf{x_t}' \mathscr{E}_\dagger(w_t^{\dagger 2}) \\
=&\frac{1}{T-p-q} \sum_{t=p+q+1}^{T} h_t \mathbf{x_t} \mathbf{x_t}' + \underbrace{\frac{1}{T-p-q} \sum_{t=p+q+1}^{T} \left(\widetilde{h}_t - h_t\right) \mathbf{x_t} \mathbf{x_t}'}_{\overset{Lemma\ 3.1}{=} o_p(1)} \\
\xrightarrow{p}&\mathbf{V}.
\end{aligned}$$

Let $\mathbf{c} \in \mathbb{R}^{p+q+1}$ and

$$s_T^2 := \sum_{t=p+q+1}^{T} \mathscr{E}_\dagger\left(\mathbf{c}' \mathbf{y_t^\dagger}\right)^2 \xrightarrow{p} \mathbf{c}' \mathbf{V} \mathbf{c},$$

then we have the Lyapounov condition for $\delta = 1$

$$\begin{aligned}
&\sum_{t=p+q+1}^{T} \frac{1}{s_T^3} \mathscr{E}_\dagger \left|\mathbf{c}' \mathbf{y_t^\dagger}\right|^3 \\
=&\frac{\mathscr{E}_\dagger |w_t^\dagger|^3}{s_T^3 (T-p-q)^{1/2}} \underbrace{\frac{1}{T-p-q} \sum_{t=p+q+1}^{T} \left|\mathbf{c}' \sqrt{\widetilde{h}_t} \mathbf{x_t}\right|^3}_{\overset{Lemma\ 3.2}{=} O_p(1)} \\
=&o_p(1).
\end{aligned}$$

The Cramér-Wold theorem and the central limit theorem for triangular arrays applied to the sequence $\mathbf{c}' \mathbf{y_1^\dagger}, ..., \mathbf{c}' \mathbf{y_T^\dagger}$ show

$$\sum_{t=p+q+1}^{T} \mathbf{y_t^\dagger} \xrightarrow{d} \mathscr{N}\left(0, \mathbf{V}\right) \quad \textit{in probability.}$$

□

Theorem 3.6

Suppose that X_t is generated by the model (3.1) satisfying assumptions (A1)-(A4) and (B1)-(B4). Then the wild bootstrap of section 3.3 for the ARCH part has the following asymptotic property:

$$\sqrt{T-p-q}(\widetilde{\mathbf{b}^\dagger}-\widetilde{\mathbf{b}}) \xrightarrow{d} \mathcal{N}\left(0, \mathbf{B}^{-1}\tau\mathbf{W}\mathbf{B}^{-1}\right) \quad \textit{in probability},$$

where $\tau = \frac{2}{\kappa-1}$.

Proof. The proof is the mirror image of Theorem 3.4. From Theorem 3.2 we already have

$$\frac{1}{T-p-q}\widehat{\mathbf{E}}'\widehat{\mathbf{E}} \xrightarrow{p} \mathbf{B}.$$

Together with the Slutsky theorem it suffices to show

$$\frac{1}{\sqrt{T-p-q}}\widehat{\mathbf{E}}'\mathbf{m}^\dagger = \frac{1}{\sqrt{T-p-q}}\sum_{t=p+q+1}^{T} m_t^\dagger \widehat{\mathbf{e}_\mathbf{t}} \xrightarrow{d} \mathcal{N}\left(0, \mathbf{W}\right) \quad \textit{in probability.}$$

Now we observe

$$\mathbf{z}_\mathbf{t}^\dagger := \frac{1}{\sqrt{T-p-q}} m_t^\dagger \widehat{\mathbf{e}_\mathbf{t}} = \frac{1}{\sqrt{T-p-q}} \widetilde{h}_t (w_t^{\dagger 2}-1)\widehat{\mathbf{e}_\mathbf{t}}, \quad t = p+q+1, \ ..., \ T.$$

For each T the sequence $\mathbf{z}_\mathbf{1}^\dagger, ..., \mathbf{z}_\mathbf{T}^\dagger$ is independent because $w_1^\dagger, ..., w_T^\dagger$ is independent. Here we obtain

$$\mathscr{E}_\dagger\left(\mathbf{z}_\mathbf{t}^\dagger\right) = \frac{1}{\sqrt{T-p-q}} \widetilde{h}_t \widehat{\mathbf{e}_\mathbf{t}} \mathscr{E}_\dagger (w_t^{\dagger 2}-1) = 0$$

and from Lemma 3.1

$$\begin{aligned}
&\sum_{t=p+q+1}^{T} \mathscr{E}_\dagger\left(\mathbf{z}_\mathbf{t}^\dagger \mathbf{z}_\mathbf{t}^{\dagger\prime}\right)\\
=&\frac{1}{T-p-q}\sum_{t=p+q+1}^{T}\left(\mathscr{E}_\dagger\left(w_t^{\dagger 4}\right) - 2\mathscr{E}_\dagger (w_t^{\dagger 2}) + 1\right)\widetilde{h}_t^2 \widehat{\mathbf{e}_\mathbf{t}}\widehat{\mathbf{e}_\mathbf{t}}'\\
=&2\left(\frac{1}{T-p-q}\sum_{t=p+q+1}^{T} h_t^2 \mathbf{e}_\mathbf{t}\mathbf{e}_\mathbf{t}' + o_p(1)\right)\\
\xrightarrow{p}& \tau\mathbf{W}.
\end{aligned}$$

Let $\mathbf{c} \in \mathbb{R}^{p+q+1}$ and

$$s_T^2 := \sum_{t=p+q+1}^{T} \mathscr{E}_{\dagger}(\mathbf{c}'\mathbf{z}_t^{\dagger})^2 \xrightarrow{p} \mathbf{c}'\tau\mathbf{W}\mathbf{c},$$

then we have the Lyapounov condition for $\delta = 1$

$$\begin{aligned}
&\sum_{t=p+q+1}^{T} \frac{1}{s_T^3}\mathscr{E}_{\dagger}\left|\mathbf{c}'\mathbf{z}_t^{\dagger}\right|^3 \\
=&\sum_{t=p+q+1}^{T} \frac{1}{s_T^3}\mathscr{E}_{\dagger}\left|\mathbf{c}'\frac{1}{\sqrt{T-p-q}}\widetilde{h}_t(w_t^{\dagger 2}-1)\widehat{\mathbf{e}}_{\mathbf{t}}\right|^3 \\
=&\frac{\mathscr{E}_{\dagger}|w_t^{\dagger 2}-1|^3}{s_T^3(T-p-q)^{1/2}}\underbrace{\frac{1}{T-p-q}\sum_{t=p+q+1}^{T}\left|\mathbf{c}'\widetilde{h}_t\widehat{\mathbf{e}}_{\mathbf{t}}\right|^3}_{\overset{Lemma\ 3.2}{=} O_p(1)} \\
=&o_p(1).
\end{aligned}$$

The Cramér-Wold theorem and the central limit theorem for triangular arrays applied to the sequence $\mathbf{c}'\mathbf{z}_1^{\dagger}, \dots, \mathbf{c}'\mathbf{z}_T^{\dagger}$ show

$$\sum_{t=p+q+1}^{T} \mathbf{z}_t^{\dagger} \xrightarrow{d} \mathscr{N}\left(0, \tau\mathbf{W}\right) \quad \textit{in probability.}$$

□

Remark 3.6

Theorem 3.6 implies that the wild bootstrap of section 3.3 is weakly consistent only if $\kappa = 3$. In other words, if $\kappa \neq 3$, it is necessary to construct a wild bootstrap in a different way so that $\mathscr{E}_{\dagger}(w_t^{\dagger}) = 0$, $\mathscr{E}_{\dagger}(w_t^{\dagger 2}) = 1$, $\mathscr{E}_{\dagger}(w_t^{\dagger 3}) \xrightarrow{p} 0$ and $\mathscr{E}_{\dagger}(w_t^{\dagger 4}) \xrightarrow{p} \kappa$; however, it is not always easy to obtain such $w_t^{\dagger}$. In the next section we will see the result of a false application, in which the wild bootstrap with standard normal distribution is applied though $\kappa = 9$.

Remark 3.7

A possible way to avoid the problem of the 'one-step' wild bootstrap is that we apply another bootstrap procedure again to the ARCH part. We can continue the 'two-step' wild bootstrap method as follows:

(step 5) Generate the bootstrap process $\varepsilon_t^{\ddagger 2}$ by computing

$$\varepsilon_t^{\ddagger 2} = \widetilde{b}_0 + \sum_{j=1}^{q} \widetilde{b}_j \widehat{\varepsilon_{t-j}}^2 + \sqrt{\kappa - 1}\widetilde{h}_t w_t^{\ddagger},$$

$$w_t^{\ddagger} \overset{iid}{\sim} \mathscr{N}(0,\, 1), \quad t = p+q+1,\, \dots,\, T,$$

or in vector and matrix notation

$$\mathbf{e}^{\ddagger} = \widehat{\mathbf{E}}\,\widetilde{\mathbf{b}} + \mathbf{n}^{\ddagger},$$

$$\begin{pmatrix} \varepsilon_{p+q+1}^{\ddagger 2} \\ \varepsilon_{p+q+2}^{\ddagger 2} \\ \vdots \\ \varepsilon_{T}^{\ddagger 2} \end{pmatrix} = \begin{pmatrix} 1 & \widehat{\varepsilon_{p+q}}^2 & \widehat{\varepsilon_{p+q-1}}^2 & \cdots & \widehat{\varepsilon_{p+1}}^2 \\ 1 & \widehat{\varepsilon_{p+q+1}}^2 & \widehat{\varepsilon_{p+q}}^2 & \cdots & \widehat{\varepsilon_{p+2}}^2 \\ \vdots & \vdots & \vdots & \ddots & \vdots \\ 1 & \widehat{\varepsilon_{T-1}}^2 & \widehat{\varepsilon_{T-2}}^2 & \cdots & \widehat{\varepsilon_{T-q}}^2 \end{pmatrix} \begin{pmatrix} \widetilde{b_0} \\ \widetilde{b_1} \\ \widetilde{b_2} \\ \vdots \\ \widetilde{b_q} \end{pmatrix} + \sqrt{\kappa-1} \begin{pmatrix} \widetilde{h_{p+q+1}} w_{p+q+1}^{\ddagger} \\ \widetilde{h_{p+q+2}} w_{p+q+2}^{\ddagger} \\ \vdots \\ \widetilde{h_T} w_T^{\ddagger} \end{pmatrix}.$$

(step 6) Calculate the bootstrap estimator

$$\widetilde{\mathbf{b}^{\ddagger}} = (\widehat{\mathbf{E}}'\widehat{\mathbf{E}})^{-1}\widehat{\mathbf{E}}'\mathbf{e}^{\ddagger}.$$

Here we observe

$$\begin{aligned} \widetilde{\mathbf{b}^{\ddagger}} &= (\widehat{\mathbf{E}}'\widehat{\mathbf{E}})^{-1}\widehat{\mathbf{E}}'(\widehat{\mathbf{E}}\widetilde{\mathbf{b}} + \mathbf{n}^{\ddagger}) \\ &= \widetilde{\mathbf{b}} + (\widehat{\mathbf{E}}'\widehat{\mathbf{E}})^{-1}\widehat{\mathbf{E}}'\mathbf{n}^{\ddagger}, \end{aligned}$$

that is,

$$\sqrt{T-p-q}(\widetilde{\mathbf{b}^{\ddagger}} - \widetilde{\mathbf{b}}) = \left(\frac{1}{T-p-q}\widehat{\mathbf{E}}'\widehat{\mathbf{E}}\right)^{-1}\left(\frac{1}{\sqrt{T-p-q}}\widehat{\mathbf{E}}'\mathbf{n}^{\ddagger}\right).$$

Theorem 3.7

Suppose that X_t is generated by the model (3.1) satisfying assumptions (A1)-(A4) and (B1)-(B4). Then the wild bootstrap of Remark 3.7 is weakly consistent, i.e.

$$\sqrt{T-p-q}(\widetilde{\mathbf{b}^{\ddagger}} - \widetilde{\mathbf{b}}) \xrightarrow{d} \mathcal{N}\left(0,\, \mathbf{B}^{-1}\mathbf{W}\mathbf{B}^{-1}\right) \quad \textit{in probability.}$$

Proof. The proof is the mirror image of Theorem 3.4. From Theorem 3.2 we already have

$$\frac{1}{T-p-q}\widehat{\mathbf{E}}'\widehat{\mathbf{E}} \xrightarrow{p} \mathbf{B}.$$

Together with the Slutsky theorem it suffices to show

$$\frac{1}{\sqrt{T-p-q}}\widehat{\mathbf{E}}'\mathbf{n}^{\ddagger} = \sqrt{\frac{\kappa-1}{T-p-q}} \sum_{t=p+q+1}^{T} \widetilde{h_t} w_t^{\ddagger} \widehat{\mathbf{e_t}} \xrightarrow{d} \mathcal{N}\left(0, \mathbf{W}\right) \quad \textit{in probability.}$$

Now we observe

$$\mathbf{z_t^{\ddagger}} := \sqrt{\frac{\kappa-1}{T-p-q}} \widetilde{h_t} w_t^{\ddagger} \widehat{\mathbf{e_t}}, \quad t = p+q+1, \ldots, T.$$

For each T the sequence $\mathbf{z_1^\ddagger},...,\mathbf{z_T^\ddagger}$ is independent because $w_1^\ddagger,...,w_T^\ddagger$ is independent. Here we obtain $\mathscr{E}_\ddagger(\mathbf{z_t^\ddagger}) = 0$ and from Lemma 3.1

$$\begin{aligned}
&\sum_{t=p+q+1}^{T} \mathscr{E}_\ddagger\left(\mathbf{z_t^\ddagger z_t^\ddagger}'\right) \\
=&\frac{\kappa-1}{T-p-q}\sum_{t=p+q+1}^{T} \widetilde{h_t}^2 \underbrace{\mathscr{E}_\ddagger\left(w_t^{\ddagger 2}\right)}_{=1} \widehat{\mathbf{e_t}}\widehat{\mathbf{e_t}}' \\
=&\frac{\kappa-1}{T-p-q}\sum_{t=p+q+1}^{T} \left(h_t^2 \mathbf{e_t e_t}' + o_p(1)\right) \\
\xrightarrow{p}&\mathbf{W}.
\end{aligned}$$

Let $\mathbf{c} \in \mathbb{R}^{p+q+1}$ and

$$s_T^2 := \sum_{t=p+q+1}^{T} \mathscr{E}_\ddagger(\mathbf{c}'\mathbf{z_t^\ddagger})^2 \xrightarrow{p} \mathbf{c}'\mathbf{W}\mathbf{c},$$

then we have the Lyapounov condition for $\delta = 1$

$$\begin{aligned}
&\sum_{t=p+q+1}^{T} \frac{1}{s_T^3}\mathscr{E}_\ddagger\left(\left|\mathbf{c}'\mathbf{z_t^\ddagger}\right|^3\right) \\
=&\sum_{t=p+q+1}^{T} \frac{1}{s_T^3}\mathscr{E}_\ddagger\left(\left|\mathbf{c}'\sqrt{\frac{\kappa-1}{T-p-q}}\widetilde{h}_t w_t^\ddagger \widehat{\mathbf{e_t}}\right)\right|^3\right) \\
=&\frac{(\kappa-1)^{3/2}}{s_T^3(T-q)^{1/2}} \underbrace{\mathscr{E}_\ddagger|w_t^\ddagger|^3}_{=O_p(1)} \underbrace{\frac{1}{T-p-q}\sum_{t=p+q+1}^{T}\left|\mathbf{c}'\widetilde{h}_t\widehat{\mathbf{e_t}}\right|^3}_{\overset{Lemma\ 3.2}{=} O_p(1)} \\
=&o_p(1).
\end{aligned}$$

The Cramér-Wold theorem and the central limit theorem for triangular arrays applied to the sequence $\mathbf{c}'\mathbf{z_1^\ddagger},...,\mathbf{c}'\mathbf{z_T^\ddagger}$ show

$$\sum_{t=p+q+1}^{T} \mathbf{z_t}^\ddagger \xrightarrow{d} \mathscr{N}\left(0, \mathbf{W}\right) \quad \textit{in probability.}$$

□

3.4 Simulations

In this section we demonstrate the large sample properties of the residual bootstrap of section 3.2 and the wild bootstrap of section 3.3. Let us consider the following parametric AR(1)-ARCH(1) model:

$$X_t = a_0 + a_1 X_{t-1} + \varepsilon_t,$$
$$\varepsilon_t = \sqrt{h_t}\eta_t,$$
$$h_t = b_0 + b_1 \varepsilon_{t-1}^2, \qquad t = 1, ..., 50,000.$$

The simulation procedure is as follows.

(step 1) Simulate the bias $\eta_t \overset{iid}{\sim} \sqrt{\frac{3}{5}}\, t_5$ for $t = 1, ..., 50,000$. Note that $\mathscr{E}(\eta_t) = \mathscr{E}(\eta_t^3) = 0$, $\mathscr{E}(\eta_t^2) = 1$ and $\mathscr{E}(\eta_t^4) = 9$.

(step 2) Let $X_0 = \mathscr{E}(X_t) = \frac{a_0}{1-a_1}$, $h_0 = \varepsilon_0^2 = \mathscr{E}(\varepsilon_t^2) = \frac{b_0}{1-b_1}$, $a_0 = 0.141$, $a_1 = 0.433$, $b_0 = 0.007$, $b_1 = 0.135$ and calculate

$$X_t = a_0 + a_1 X_{t-1} + \left(b_0 + b_1 \varepsilon_{t-1}^2\right)^{1/2} \eta_t$$

for $t = 1, ..., 50,000$.

(step 3) From the data set $\{X_1, X_2, ..., X_{50,000}\}$ obtain the OLS estimators $\widehat{a_0}$, $\widehat{a_1}$, $\widetilde{b_0}$ and $\widetilde{b_1}$.

(step 4) Repeat (step 1)-(step 3) 2,000 times.

(step 5) Choose one data set $\{X_1, X_2, ..., X_{50,000}\}$ and its estimators $\{\widehat{a_0}, \widehat{a_1}, \widetilde{b_0}, \widetilde{b_1}\}$. Adopt the residual bootstrap of section 3.2 and the wild bootstrap of section 3.3 based on the chosen data set and estimators. Both bootstrap methods are repeated 2,000 times.

(step 6) Compare the simulated density of the OLS estimators ($\sqrt{T}(\widehat{a_0} - a_0)$ etc., thin lines) with the residual and the wild bootstrap approximations ($\sqrt{T}(\widehat{a_0^*} - \widehat{a_0})$, $\sqrt{T}(\widehat{a_0^\dagger} - \widehat{a_0})$ etc., bold lines).

Figure 3.1 illustrates that the residual bootstrap is weakly consistent. Moreover, Figure 3.1 (c)-(d) indicates that in finite samples the residual bootstrap yields more accurate approximation than the conventional normal approximation. The theoretical explanation of the result is leaved for further research.

On the other hand, Figure 3.2 (c)-(d) shows the problem of the wild bootstrap for the ARCH part, which corresponds to the theoretical result in Theorem 3.6 with $\tau = \frac{1}{4}$.

Figure 3.3 also shows the problem of the two-step wild bootstrap, which seems to contradict Theorem 3.7. In fact, with $T = 50,000$ the approximation error

$$\frac{1}{T}\sum_{t=1}^{T}\eta_t^4 - \mathscr{E}(\eta_t^4)$$

is not negligible, which results in the large error of the bootstrap approximation. That is, although the two-step wild bootstrap works theoretically, it is of limited use for the AR(1)-ARCH(1) model.

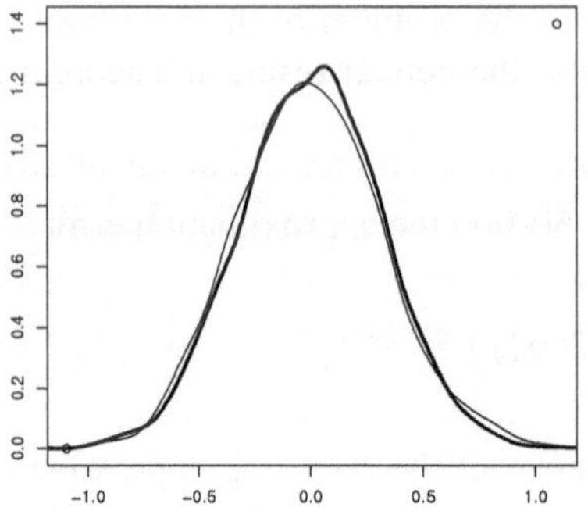

(a) $\sqrt{T}(\widehat{a_0} - a_0)$ and $\sqrt{T}(\widehat{a_0^*} - \widehat{a_0})$

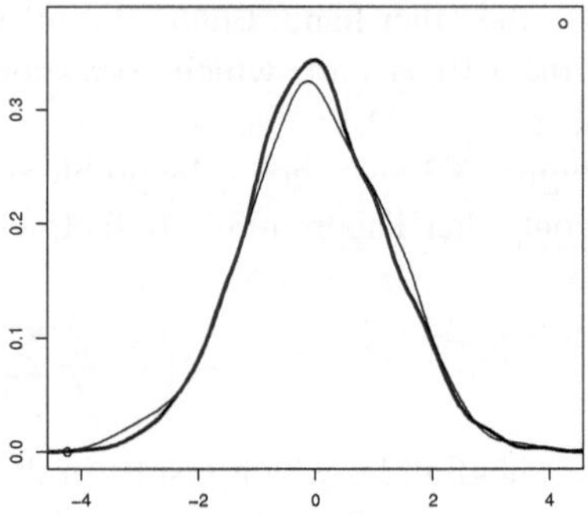

(b) $\sqrt{T}(\widehat{a_1} - a_1)$ and $\sqrt{T}(\widehat{a_1^*} - \widehat{a_1})$

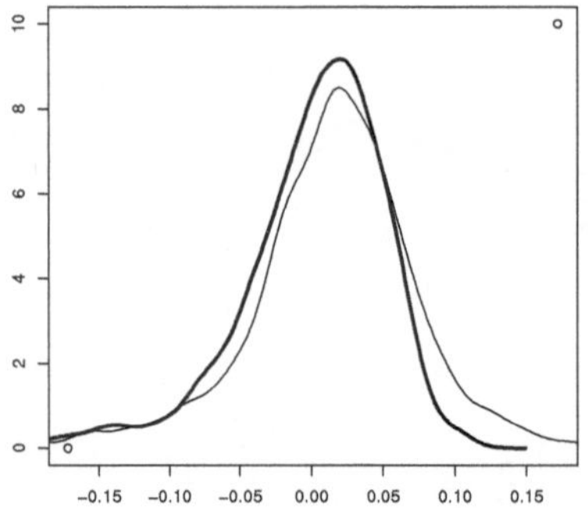

(c) $\sqrt{T}(\widetilde{b_0} - b_0)$ and $\sqrt{T}(\widetilde{b_0^*} - \widetilde{b_0})$

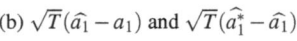

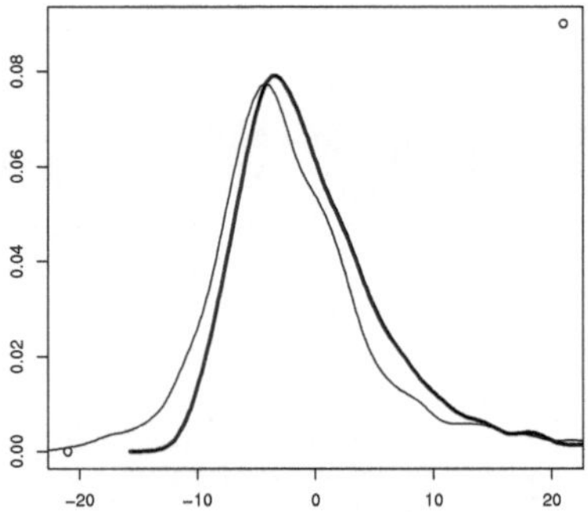

(d) $\sqrt{T}(\widetilde{b_1} - b_1)$ and $\sqrt{T}(\widetilde{b_1^*} - \widetilde{b_1})$

Figure 3.1: Residual Bootstrap

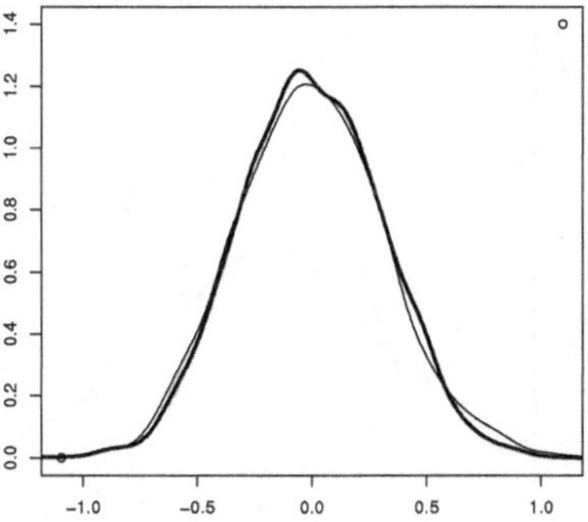

(a) $\sqrt{T}(\widehat{a_0} - a_0)$ and $\sqrt{T}(\widehat{a_0^\dagger} - \widehat{a_0})$

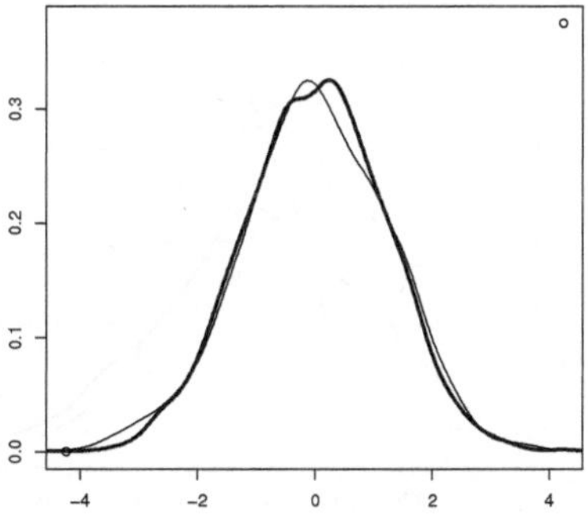

(b) $\sqrt{T}(\widehat{a_1} - a_1)$ and $\sqrt{T}(\widehat{a_1^\dagger} - \widehat{a_1})$

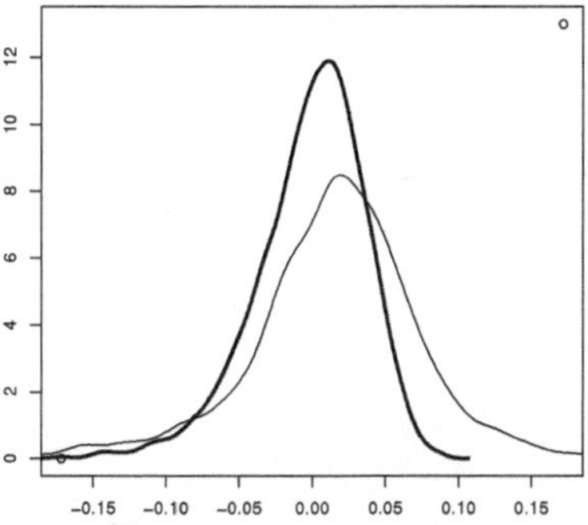

(c) $\sqrt{T}(\widetilde{b_0} - b_0)$ and $\sqrt{T}(\widetilde{b_0^\dagger} - \widetilde{b_0})$

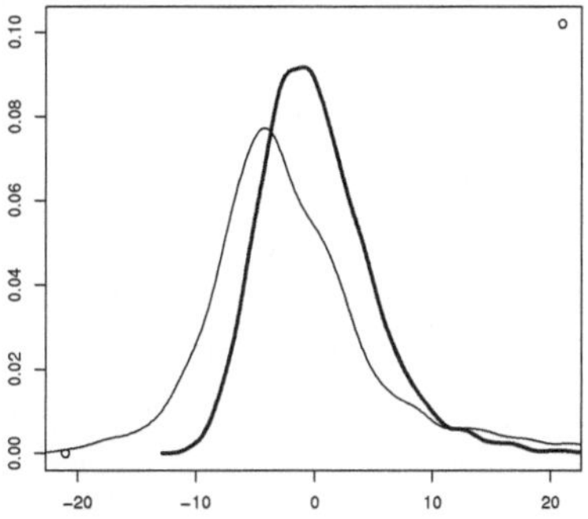

(d) $\sqrt{T}(\widetilde{b_1} - b_1)$ and $\sqrt{T}(\widetilde{b_1^\dagger} - \widetilde{b_1})$

Figure 3.2: Wild Bootstrap

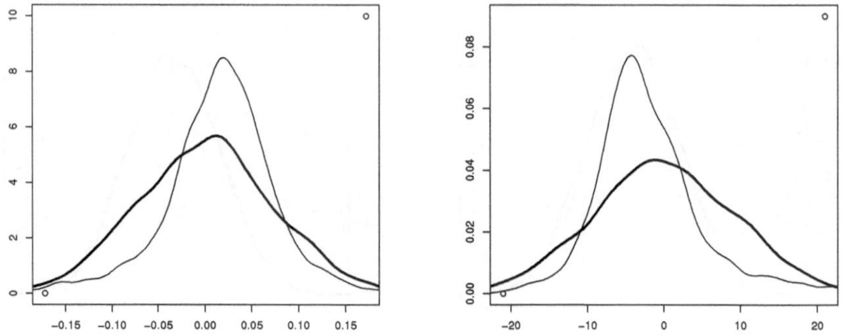

(a) $\sqrt{T}(\widetilde{b_0} - b_0)$ and $\sqrt{T}(\widetilde{b_0^{\ddagger}} - \widetilde{b_0})$ (b) $\sqrt{T}(\widetilde{b_1} - b_1)$ and $\sqrt{T}(\widetilde{b_1^{\ddagger}} - \widetilde{b_1})$

Figure 3.3: Two-step Wild Bootstrap

4 Parametric ARMA(p, q)- GARCH(r, s) Models

In this chapter we extend the results of the previous chapter to the parametric ARMA(p, q)-GARCH(r, s) model estimated by the QML method. In the first section we sketch the estimation theory based on Francq and Zakoïan (2004). Then, analogously to the previous chapter, possible applications of the residual and the wild bootstrap are proposed and their weak consistency investigated. These theoretical results are confirmed by simulations in the last section.

4.1 Estimation Theory

Since Bollerslev (1986) introduced the GARCH model, asymptotic theories for GARCH-type models have been presented by many researchers. Lee and Hansen (1994) and Lumsdaine (1996) examined asymptotic properties of the QML estimators for the GARCH(1, 1) model. Properties of the GARCH(p, q) model have been investigated by Berkes, Horváth and Kokoszka (2003) and Comte and Lieberman (2003). For the more general ARMA(p, q)-GARCH(r, s) model Ling and Li (1997) proved consistency and asymptotic normality of the local QML estimators. Properties of the global QML were investigated by Francq and Zakoïan (2004). See also Ling and McAleer (2003a) for vector ARMA-GARCH models, Ling and McAleer (2003b) for nonstationary ARMA-GARCH models, and Ling (2007) for self-weighted QML estimators for ARMA-GARCH models. In this section we introduce the ARMA(p, q)-GARCH(r, s) model and observe the asymptotic properties of the QML estimators based on Francq and Zakoïan (2004).

4.1.1 Model and Assumptions

Assume that $\{X_t,\ t = 1,\ ...,\ T\}$ are generated by the following ARMA(p, q)-GARCH(r, s) model:

$$\begin{aligned} X_t &= a_0 + \sum_{i=1}^{p} a_i X_{t-i} + \sum_{j=1}^{q} \alpha_j \varepsilon_{t-j} + \varepsilon_t, \\ \varepsilon_t &= \sqrt{h_t}\eta_t, \\ h_t &= b_0 + \sum_{j=1}^{s} b_j \varepsilon_{t-j}^2 + \sum_{i=1}^{r} \beta_i h_{t-i}, \end{aligned} \tag{4.1}$$

where $b_0 > 0$, $b_j \geq 0, j = 1,...,s$, $\beta_i \geq 0, i = 1,...,r$, and $\{\eta_t\}$ is a sequence of i.i.d. random variables such that $\mathscr{E}(\eta_t) = 0$, $\mathscr{E}(\eta_t^2) = 1$, following a symmetric distribution, i.e. $\mathscr{E}(\eta_t^3) = 0$, and $\mathscr{E}(\eta_t^4) =: \kappa < \infty$. The parameter space $\Phi = \Phi_{\mathbf{a}} \times \Phi_{\mathbf{b}}$ is compact, where $\Phi_{\mathbf{a}} \subset R^{p+q+1}$ and $\Phi_{\mathbf{b}} \subset R_0^{s+r+1}$ with $R = (-\infty, \infty)$ and $R_0 = [0, \infty)$. Theoretically, ε_t and h_t depend on the infinite past history of $\{X_t\}$ or $\{\varepsilon_t\}$. For simplicity, however, we assume here the presample values of X_t, ε_t and h_t, which does not affect the asymptotic results (see Bollerslev (1986)).

Let $\boldsymbol{\theta}_0 = (\mathbf{a_0}', \mathbf{b_0}')'$, $\mathbf{a_0} = (a_{00}\ a_{01}\ ...\ a_{0p}\ \alpha_{01}\ ...\ \alpha_{0q})'$ and $\mathbf{b_0} = (b_{00}\ b_{01}\ ...\ b_{0s}\ \beta_{01}\ ...\ \beta_{0r})'$ be the true parameter vector. Denote $\boldsymbol{\theta} = (\mathbf{a}', \mathbf{b}')'$, $\mathbf{a} = (a_0\ a_1\ ...\ a_p\ \alpha_1\ ...\ \alpha_q)'$ and $\mathbf{b} = (b_0\ b_1\ ...\ b_s\ \beta_1\ ...\ \beta_r)'$, and

$$\mathscr{A}_{\mathbf{a}}(z) = 1 - \sum_{i=1}^{p} a_i z^i, \qquad \mathscr{B}_{\mathbf{a}}(z) = 1 + \sum_{j=1}^{q} \alpha_j z^j,$$

$$\mathscr{A}_{\mathbf{b}}(z) = \sum_{j=1}^{s} b_j z^j \quad \textit{and} \quad \mathscr{B}_{\mathbf{b}}(z) = 1 - \sum_{i=1}^{r} \beta_i z^i.$$

We further assume that

(A1) η_t^2 has a non-degenerate distribution.

(A2) If $r > 0$, $\mathscr{A}_{\mathbf{b_0}}(z)$ and $\mathscr{B}_{\mathbf{b_0}}(z)$ have no common root, $\mathscr{A}_{\mathbf{b_0}}(1) \neq 0$, $b_{0s} + \beta_{0r} \neq 0$ and $\sum_{i=1}^{r} \beta_i < 1$ for all $\mathbf{b} \in \Phi_{\mathbf{b}}$.

(A3) For all $\boldsymbol{\theta} \in \Phi$ $\mathscr{A}_{\mathbf{a}}(z)\mathscr{B}_{\mathbf{a}}(z) = 0$ implies $|z| > 1$.

(A4) $\mathscr{A}_{\mathbf{a_0}}(z)$ and $\mathscr{B}_{\mathbf{a_0}}(z)$ have no common root, $a_{0p} \neq 0$ or $\alpha_{0q} \neq 0$.

(A5) $\boldsymbol{\theta}_0 \in \mathring{\Phi}$, where $\mathring{\Phi}$ denotes the interior of Φ.

4.1.2 QML Estimation

The log likelihood function of the model (4.1) for a sample of T observations is

$$L_T(\boldsymbol{\theta}) = \frac{1}{T}\sum_{t=1}^{T} l_t(\boldsymbol{\theta}) \quad with \quad l_t(\boldsymbol{\theta}) = -\frac{1}{2}\log h_t(\boldsymbol{\theta}) - \frac{\varepsilon_t^2}{2h_t}(\boldsymbol{\theta}).$$

A QML estimator $\widehat{\boldsymbol{\theta}}$ is any measurable solution of

$$\widehat{\boldsymbol{\theta}} = \arg\max_{\boldsymbol{\theta}\in\Phi} L_T(\boldsymbol{\theta}).$$

Theorem 4.1

Suppose that X_t is generated by the model (4.1) satisfying assumptions (A1)-(A5). Then

$$\sqrt{T}(\widehat{\boldsymbol{\theta}} - \boldsymbol{\theta}_0) \xrightarrow{d} \mathcal{N}\left(0, \boldsymbol{\Sigma}^{-1}\boldsymbol{\Omega}\boldsymbol{\Sigma}^{-1}\right),$$

where

$$\boldsymbol{\Sigma} = \mathscr{E}\left(\frac{\partial^2 l_t(\boldsymbol{\theta}_0)}{\partial\boldsymbol{\theta}\partial\boldsymbol{\theta}'}\right) = \begin{pmatrix} \boldsymbol{\Sigma}_1 & 0 \\ 0 & \boldsymbol{\Sigma}_2 \end{pmatrix},$$

$$\boldsymbol{\Omega} = \mathscr{E}\left(\frac{\partial l_t(\boldsymbol{\theta}_0)}{\partial\boldsymbol{\theta}}\frac{\partial l_t(\boldsymbol{\theta}_0)}{\partial\boldsymbol{\theta}'}\right) = \begin{pmatrix} \boldsymbol{\Omega}_1 & 0 \\ 0 & \boldsymbol{\Omega}_2 \end{pmatrix}$$

with

$$\boldsymbol{\Sigma}_1 = \mathscr{E}\left(\frac{1}{2h_t^2(\boldsymbol{\theta}_0)}\frac{\partial h_t(\boldsymbol{\theta}_0)}{\partial\mathbf{a}}\frac{\partial h_t(\boldsymbol{\theta}_0)}{\partial\mathbf{a}'}\right) + \mathscr{E}\left(\frac{1}{h_t(\boldsymbol{\theta}_0)}\frac{\partial\varepsilon_t(\boldsymbol{\theta}_0)}{\partial\mathbf{a}}\frac{\partial\varepsilon_t(\boldsymbol{\theta}_0)}{\partial\mathbf{a}'}\right),$$

$$\boldsymbol{\Sigma}_2 = \mathscr{E}\left(\frac{1}{2h_t^2(\boldsymbol{\theta}_0)}\frac{\partial h_t(\boldsymbol{\theta}_0)}{\partial\mathbf{b}}\frac{\partial h_t(\boldsymbol{\theta}_0)}{\partial\mathbf{b}'}\right),$$

$$\boldsymbol{\Omega}_1 = \frac{\kappa-1}{2}\mathscr{E}\left(\frac{1}{2h_t^2(\boldsymbol{\theta}_0)}\frac{\partial h_t(\boldsymbol{\theta}_0)}{\partial\mathbf{a}}\frac{\partial h_t(\boldsymbol{\theta}_0)}{\partial\mathbf{a}'}\right) + \mathscr{E}\left(\frac{1}{h_t(\boldsymbol{\theta}_0)}\frac{\partial\varepsilon_t(\boldsymbol{\theta}_0)}{\partial\mathbf{a}}\frac{\partial\varepsilon_t(\boldsymbol{\theta}_0)}{\partial\mathbf{a}'}\right),$$

$$\boldsymbol{\Omega}_2 = \frac{\kappa-1}{2}\mathscr{E}\left(\frac{1}{2h_t^2(\boldsymbol{\theta}_0)}\frac{\partial h_t(\boldsymbol{\theta}_0)}{\partial\mathbf{b}}\frac{\partial h_t(\boldsymbol{\theta}_0)}{\partial\mathbf{b}'}\right).$$

Proof. Francq and Zakoïan (2004), Theorem 3.2. □

Theorem 4.2

Suppose that X_t is generated by the model (4.1) satisfying assumptions (A1)-

(A5). Then the matrices $\boldsymbol{\Sigma}$ and $\boldsymbol{\Omega}$ in Theorem 4.1 can be consistently estimated by

$$\widehat{\boldsymbol{\Sigma}_1} = \frac{1}{T}\sum_{t=1}^{T}\left(\frac{1}{2h_t^2(\widehat{\boldsymbol{\theta}})}\frac{\partial h_t(\widehat{\boldsymbol{\theta}})}{\partial \mathbf{a}}\frac{\partial h_t(\widehat{\boldsymbol{\theta}})}{\partial \mathbf{a}'}\right) + \frac{1}{T}\sum_{t=1}^{T}\left(\frac{1}{h_t(\widehat{\boldsymbol{\theta}})}\frac{\partial \varepsilon_t(\widehat{\boldsymbol{\theta}})}{\partial \mathbf{a}}\frac{\partial \varepsilon_t(\widehat{\boldsymbol{\theta}})}{\partial \mathbf{a}'}\right),$$

$$\widehat{\boldsymbol{\Sigma}_2} = \frac{1}{T}\sum_{t=1}^{T}\left(\frac{1}{2h_t^2(\widehat{\boldsymbol{\theta}})}\frac{\partial h_t(\widehat{\boldsymbol{\theta}})}{\partial \mathbf{b}}\frac{\partial h_t(\widehat{\boldsymbol{\theta}})}{\partial \mathbf{b}'}\right),$$

$$\widehat{\boldsymbol{\Omega}_1} = \frac{1}{2}\left(\frac{1}{T}\sum_{t=1}^{T}\frac{\varepsilon_t^4}{h_t^2}(\widehat{\boldsymbol{\theta}}) - 1\right)\frac{1}{T}\sum_{t=1}^{T}\left(\frac{1}{2h_t^2(\widehat{\boldsymbol{\theta}})}\frac{\partial h_t(\widehat{\boldsymbol{\theta}})}{\partial \mathbf{a}}\frac{\partial h_t(\widehat{\boldsymbol{\theta}})}{\partial \mathbf{a}'}\right)$$
$$+\frac{1}{T}\sum_{t=1}^{T}\left(\frac{1}{h_t(\widehat{\boldsymbol{\theta}})}\frac{\partial \varepsilon_t(\widehat{\boldsymbol{\theta}})}{\partial \mathbf{a}}\frac{\partial \varepsilon_t(\widehat{\boldsymbol{\theta}})}{\partial \mathbf{a}'}\right),$$

$$\widehat{\boldsymbol{\Omega}_2} = \frac{1}{2}\left(\frac{1}{T}\sum_{t=1}^{T}\frac{\varepsilon_t^4}{h_t^2}(\widehat{\boldsymbol{\theta}}) - 1\right)\frac{1}{T}\sum_{t=1}^{T}\left(\frac{1}{2h_t^2(\widehat{\boldsymbol{\theta}})}\frac{\partial h_t(\widehat{\boldsymbol{\theta}})}{\partial \mathbf{b}}\frac{\partial h_t(\widehat{\boldsymbol{\theta}})}{\partial \mathbf{b}'}\right).$$

Proof. Ling (2007, p. 854). Although Ling (2007) discussed the self-weighted QML estimator for the ARMA(p, q)-GARCH(r, s) model, we can analogously prove Theorem 4.2 assuming that the weight $w_t = 1$. See also Theorem 5.1 in Ling and McAleer (2003a). □

4.2 Residual Bootstrap

There have been numbers of works introducing bootstrap methods for ARMA or GARCH models. Kreiß and Franke (1992) proved asymptotic validity of the autoregression bootstrap technique applied to M estimators for the ARMA (p, q) model. Maercker (1997) showed weak consistency of the wild bootstrap for the GARCH(1, 1) model based on the asymptotic properties of the QML estimators that Lee and Hansen (1994) examined. Moreover, there are numbers of empirical applications of bootstrap methods for GARCH models, such as Reeves (2005), Pascual, Romo and Ruiz (2005), and Robio (1999). Despite the large number of empirical studies, there is a rather sparse literature investigating the validity of bootstrap techniques for ARMA-GARCH models.[1] In this and the following section we introduce possible ways to bootstrap the stationary ARMA(p, q)-GARCH(r, s) model based on the results of Francq and Zakoïan (2004).

[1] Reeves (2005) showed weak consistency of the bootstrap prediction for the AR(1)-GARCH(1, 1) model assuming that the autoregression bootstrap estimators for the model are weakly consistent. As far as we know, however, weak consistency of the autoregression bootstrap for the AR-GARCH model has not been proved yet.

In this section a residual bootstrap technique is proposed and its weak consistency proved. A residual bootstrap method can be applied to the model (4.1) as follows:

(step 1) Obtain the QML estimator $\widehat{\boldsymbol{\theta}}$ and calculate the residuals

$$\widehat{\varepsilon_t} = X_t - \widehat{a_0} - \sum_{i=1}^{p} \widehat{a_i} X_{t-i} - \sum_{j=1}^{q} \widehat{\alpha_j} \widehat{\varepsilon_{t-j}}, \quad t = 1, \ldots, T.$$

(step 2) Compute the estimated heteroscedasticity

$$\widehat{h_t} = \widehat{b_0} + \sum_{j=1}^{s} \widehat{b_j} \widehat{\varepsilon_{t-j}}^2 + \sum_{i=1}^{r} \widehat{\beta_i} \widehat{h_{t-i}}, \quad t = 1, \ldots, T.$$

(step 3) Compute the estimated bias

$$\widehat{\eta_t} = \frac{\widehat{\varepsilon_t}}{\sqrt{\widehat{h_t}}}$$

for $t = 1, \ldots, T$ and the standardised estimated bias

$$\widetilde{\eta_t} = \frac{\widehat{\eta_t} - \widehat{\mu}}{\widehat{\sigma}}$$

for $t = 1, \ldots, T$, where

$$\widehat{\mu} = \frac{1}{T} \sum_{t=1}^{T} \widehat{\eta_t} \quad \text{and} \quad \widehat{\sigma}^2 = \frac{1}{T} \sum_{t=1}^{T} (\widehat{\eta_t} - \widehat{\mu})^2.$$

(step 4) Obtain the empirical distribution function $\mathscr{F}_T(x)$ based on $\widetilde{\eta_t}$ defined by

$$\mathscr{F}_T(x) := \frac{1}{T} \sum_{t=1}^{T} \mathbf{1}(\widetilde{\eta_t} \leq x).$$

(step 5) Generate the bootstrap process X_t^* by computing

$$X_t^* = \widehat{a_0} + \sum_{i=1}^{p} \widehat{a_i} X_{t-i} + \sum_{j=1}^{q} \widehat{\alpha_j} \widehat{\varepsilon_{t-j}} + \varepsilon_t^*,$$

$$\varepsilon_t^* = \sqrt{\widehat{h_t}} \eta_t^*, \quad \eta_t^* \overset{iid}{\sim} \mathscr{F}_T(x), \quad t = 1, \ldots, T.$$

(step 6) Obtain the bootstrap estimator $\widehat{\mathbf{a}}^*$ by the Newton-Raphson method

$$\widehat{\mathbf{a}}^* = \widehat{\mathbf{a}} - \widehat{\mathbf{H}_1^*}(\widehat{\boldsymbol{\theta}})^{-1}\mathbf{g}_1^*(\widehat{\boldsymbol{\theta}}),$$

where

$$\mathbf{g}_1^*(\widehat{\boldsymbol{\theta}}) = \frac{1}{T}\sum_{t=1}^{T}\frac{\partial l_t^*(\widehat{\boldsymbol{\theta}})}{\partial \mathbf{a}}$$

with

$$\begin{aligned}\frac{\partial l_t^*(\widehat{\boldsymbol{\theta}})}{\partial \mathbf{a}} &= \frac{1}{2h_t(\widehat{\boldsymbol{\theta}})}\left(\frac{\varepsilon_t^{*2}}{h_t}(\widehat{\boldsymbol{\theta}}) - 1\right)\frac{\partial h_t(\widehat{\boldsymbol{\theta}})}{\partial \mathbf{a}} - \frac{\varepsilon_t^*}{h_t}(\widehat{\boldsymbol{\theta}})\frac{\partial \varepsilon_t^*(\widehat{\boldsymbol{\theta}})}{\partial \mathbf{a}} \\ &= \frac{\eta_t^{*2} - 1}{2h_t(\widehat{\boldsymbol{\theta}})}\frac{\partial h_t(\widehat{\boldsymbol{\theta}})}{\partial \mathbf{a}} - \frac{\eta_t^*}{\sqrt{h_t}(\widehat{\boldsymbol{\theta}})}\frac{\partial \varepsilon_t^*(\widehat{\boldsymbol{\theta}})}{\partial \mathbf{a}}\end{aligned}$$

and

$$\widehat{\mathbf{H}_1^*}(\widehat{\boldsymbol{\theta}}) = \frac{1}{T}\sum_{t=1}^{T}\left(\frac{1}{2h_t^2(\widehat{\boldsymbol{\theta}})}\frac{\partial h_t(\widehat{\boldsymbol{\theta}})}{\partial \mathbf{a}}\frac{\partial h_t(\widehat{\boldsymbol{\theta}})}{\partial \mathbf{a}'} + \frac{1}{h_t(\widehat{\boldsymbol{\theta}})}\frac{\partial \varepsilon_t^*(\widehat{\boldsymbol{\theta}})}{\partial \mathbf{a}}\frac{\partial \varepsilon_t^*(\widehat{\boldsymbol{\theta}})}{\partial \mathbf{a}'}\right).$$

(step 7) Analogously, calculate the bootstrap estimator $\widehat{\mathbf{b}}^*$

$$\widehat{\mathbf{b}}^* = \widehat{\mathbf{b}} - \widehat{\mathbf{H}_2^*}(\widehat{\boldsymbol{\theta}})^{-1}\mathbf{g}_2^*(\widehat{\boldsymbol{\theta}}),$$

where

$$\mathbf{g}_2^*(\widehat{\boldsymbol{\theta}}) = \frac{1}{T}\sum_{t=1}^{T}\frac{\partial l_t^*(\widehat{\boldsymbol{\theta}})}{\partial \mathbf{b}}$$

with

$$\begin{aligned}\frac{\partial l_t^*(\widehat{\boldsymbol{\theta}})}{\partial \mathbf{b}} &= \frac{1}{2h_t(\widehat{\boldsymbol{\theta}})}\left(\frac{\varepsilon_t^{*2}}{h_t}(\widehat{\boldsymbol{\theta}}) - 1\right)\frac{\partial h_t(\widehat{\boldsymbol{\theta}})}{\partial \mathbf{b}} \\ &= \frac{\eta_t^{*2} - 1}{2h_t(\widehat{\boldsymbol{\theta}})}\frac{\partial h_t(\widehat{\boldsymbol{\theta}})}{\partial \mathbf{b}}\end{aligned}$$

and

$$\widehat{\mathbf{H}_2^*}(\widehat{\boldsymbol{\theta}}) = \frac{1}{T}\sum_{t=1}^{T}\left(\frac{1}{2h_t^2(\widehat{\boldsymbol{\theta}})}\frac{\partial h_t(\widehat{\boldsymbol{\theta}})}{\partial \mathbf{b}}\frac{\partial h_t(\widehat{\boldsymbol{\theta}})}{\partial \mathbf{b}'}\right).$$

Remark 4.1

Analogously to Remark 3.4, we obtain $\mathscr{E}_*(\eta_t^*) = 0$, $\mathscr{E}_*(\eta_t^{*2}) = 1$ and $\mathscr{E}_*(\eta_t^{*k}) = \mathscr{E}(\eta_t^k) + o_p(1)$ for $k = 3, 4, \ldots$.

Remark 4.2

Here it is possible to build the bootstrap model

$$X_t^* = a_0 + \sum_{i=1}^{p} a_i X_{t-i} + \sum_{j=1}^{q} \alpha_j \varepsilon_{t-j} + \varepsilon_t^*,$$

$$\varepsilon_t = \sqrt{h_t}\eta_t, \quad \varepsilon_t^* = \sqrt{h_t}\eta_t^*, \quad \eta_t^* \overset{iid}{\sim} \mathscr{F}_T(x),$$

$$h_t = b_0 + \sum_{j=1}^{s} b_j \varepsilon_{t-j}^2 + \sum_{i=1}^{r} \beta_i h_{t-i}$$

and obtain the QML estimator $\widetilde{\boldsymbol{\theta}^*}$ by the Newton-Raphson method

$$\widetilde{\boldsymbol{\theta}^*} = \widehat{\boldsymbol{\theta}} - \mathbf{H}^*(\widehat{\boldsymbol{\theta}})^{-1}\mathbf{g}^*(\widehat{\boldsymbol{\theta}}),$$

where

$$\mathbf{g}^*(\widehat{\boldsymbol{\theta}}) = \frac{\partial L_T^*(\widehat{\boldsymbol{\theta}})}{\partial \boldsymbol{\theta}} = \frac{1}{T}\sum_{t=1}^{T} \frac{\partial l_t^*(\widehat{\boldsymbol{\theta}})}{\partial \boldsymbol{\theta}} = \left(\mathbf{g}_1^{*\prime}, \mathbf{g}_2^{*\prime}\right)'(\widehat{\boldsymbol{\theta}}),$$

$$\mathbf{H}^*(\widehat{\boldsymbol{\theta}}) = \frac{\partial^2 L_T^*(\widehat{\boldsymbol{\theta}})}{\partial \boldsymbol{\theta}\partial \boldsymbol{\theta}'} = \frac{1}{T}\sum_{t=1}^{T} \frac{\partial^2 l_t^*(\widehat{\boldsymbol{\theta}})}{\partial \boldsymbol{\theta}\partial \boldsymbol{\theta}'}.$$

However, we use $\widehat{\mathbf{H}_1^*}$ and $\widehat{\mathbf{H}_2^*}$ instead of $\mathbf{H}^*(\widehat{\boldsymbol{\theta}})$ for simplicity of the proof. Maercker (1997) also made use of this technique to bootstrap the GARCH(1,1) model.

To prove weak consistency of the residual bootstrap, we make use of the following result of Francq and Zakoïan (2004).

Lemma 4.1

Suppose that X_t is generated by the model (4.1) satisfying assumptions (A1)-(A5). Let ϕ_i denote the i-th element of vector $\mathbf{a}$ and ψ_j the j-th of $\mathbf{b}$, then for all $i = 1, \ldots, p+q+1$ and $j = 1, \ldots, r+s+1$,

$$(i) \quad \frac{1}{T}\sum_{t=1}^{T} \left|\frac{\partial \varepsilon_t(\widehat{\boldsymbol{\theta}})}{\partial \phi_i}\right|^3 = O_p(1),$$

$$(ii) \quad \frac{1}{T}\sum_{t=1}^{T} \left|\frac{\partial h_t(\widehat{\boldsymbol{\theta}})}{\partial \phi_i}\right|^3 = O_p(1),$$

$$(iii) \quad \frac{1}{T}\sum_{t=1}^{T} \left|\frac{\partial h_t(\widehat{\boldsymbol{\theta}})}{\partial \psi_j}\right|^3 = O_p(1).$$

Proof. In Francq and Zakoïan's (2004) proof of Theorem 3.2 we obtain

$$\left\| \sup_{\boldsymbol{\theta} \in \mathscr{U}(\boldsymbol{\theta}_0)} \left| \frac{\partial \varepsilon_t(\boldsymbol{\theta})}{\partial \mathbf{a}} \right| \right\|_4 < \infty,$$

$$\left\| \sup_{\boldsymbol{\theta} \in \mathscr{U}(\boldsymbol{\theta}_0)} \left| \frac{1}{h_t(\boldsymbol{\theta})} \frac{\partial h_t(\boldsymbol{\theta})}{\partial \mathbf{a}} \right| \right\|_4 < \infty,$$

$$\left\| \sup_{\boldsymbol{\theta} \in \mathscr{U}(\boldsymbol{\theta}_0)} \left| \frac{1}{h_t(\boldsymbol{\theta})} \frac{\partial h_t(\boldsymbol{\theta})}{\partial \mathbf{b}} \right| \right\|_4 < \infty$$

for any neighbourhood $\mathscr{U}(\boldsymbol{\theta}_0) \in \Phi$. Since $\frac{\partial \varepsilon_t}{\partial \phi_i}$, $\frac{\partial h_t}{\partial \phi_i}$ and $\frac{\partial h_t}{\partial \psi_j}$ are continuous at $\boldsymbol{\theta}_0$ for all $i = 1, ..., p+q+1$, $j = 1, ..., r+s+1$ and $\widehat{\boldsymbol{\theta}} \xrightarrow{p} \boldsymbol{\theta}_0$, we obtain the desired result. □

For a proof of weak consistency of the residual bootstrap we need a stronger condition than just to prove asymptotic normality of the QML estimators.

(B1) $\mathscr{E}|\eta_t|^6 < \infty$.

Theorem 4.3

Suppose that X_t is generated by the model (4.1) satisfying assumptions (A1)-(A5) and (B1). Then the residual bootstrap of section 4.2 is weakly consistent, i.e.

$$\sqrt{T}(\widehat{\boldsymbol{\theta}^*} - \widehat{\boldsymbol{\theta}}) \xrightarrow{d} \mathscr{N}\left(0, \boldsymbol{\Sigma}^{-1}\boldsymbol{\Omega}\boldsymbol{\Sigma}^{-1}\right) \quad \textit{in probability,}$$

where $\widehat{\boldsymbol{\theta}^*} = \left(\widehat{\mathbf{a}^*}', \widehat{\mathbf{b}^*}'\right)'$.

Proof. Firstly, we prove weak consistency of the GARCH part. Observe

$$\sqrt{T}(\widehat{\mathbf{b}^*} - \widehat{\mathbf{b}}) = -\widehat{\mathbf{H}_2^*}(\widehat{\boldsymbol{\theta}})^{-1} \frac{1}{\sqrt{T}} \sum_{t=1}^{T} \frac{\partial l_t^*(\widehat{\boldsymbol{\theta}})}{\partial \mathbf{b}}.$$

Note that $\widehat{\mathbf{H}_2^*}(\widehat{\boldsymbol{\theta}}) = \widehat{\boldsymbol{\Sigma}_2}$ and from Theorem 4.2 we already have $\widehat{\boldsymbol{\Sigma}_2} \xrightarrow{p} \boldsymbol{\Sigma}_2$. Together with the Slutsky theorem it is sufficient to show

$$\frac{1}{\sqrt{T}} \sum_{t=1}^{T} \frac{\partial l_t^*(\widehat{\boldsymbol{\theta}})}{\partial \mathbf{b}} \xrightarrow{d} \mathscr{N}\left(0, \boldsymbol{\Omega}_2\right) \quad \textit{in probability.}$$

Now we observe

$$\begin{aligned}\mathbf{z}_\mathbf{t}^* :=& \frac{1}{\sqrt{T}}\frac{\partial l_t^*(\widehat{\boldsymbol{\theta}})}{\partial \mathbf{b}}\\ =& \frac{1}{\sqrt{T}}\frac{\eta_t^{*2}-1}{2h_t(\widehat{\boldsymbol{\theta}})}\frac{\partial h_t(\widehat{\boldsymbol{\theta}})}{\partial \mathbf{b}}, \qquad t=1,...,T.\end{aligned}$$

For each T the sequence $\mathbf{z}_\mathbf{1}^*,...,\mathbf{z}_\mathbf{T}^*$ is independent because $\eta_1^*,...,\eta_T^*$ is independent. Here we obtain

$$\mathscr{E}_*(\mathbf{z}_\mathbf{t}^*) = \frac{1}{\sqrt{T}}\frac{1}{2h_t(\widehat{\boldsymbol{\theta}})}\frac{\partial h_t(\widehat{\boldsymbol{\theta}})}{\partial \mathbf{b}}\mathscr{E}_*\left(\eta_t^{*2}-1\right) = 0$$

and

$$\begin{aligned}&\sum_{t=1}^T \mathscr{E}_*\left(\mathbf{z}_\mathbf{t}^*\mathbf{z}_\mathbf{t}^{*\prime}\right)\\ =&\left(\mathscr{E}_*(\eta_t^{*4})-2\mathscr{E}_*(\eta_t^{*2})+1\right)\frac{1}{T}\sum_{t=1}^T\left(\frac{1}{4h_t^2(\widehat{\boldsymbol{\theta}})}\frac{\partial h_t(\widehat{\boldsymbol{\theta}})}{\partial \mathbf{b}}\frac{\partial h_t(\widehat{\boldsymbol{\theta}})}{\partial \mathbf{b}'}\right)\\ &\xrightarrow{p}\boldsymbol{\Omega}_\mathbf{2}.\end{aligned}$$

Let $\mathbf{c}\in\mathbb{R}^{s+r+1}$,

$$s_T^2 := \sum_{t=1}^T \mathscr{E}_*\left(\mathbf{c}'\mathbf{z}_\mathbf{t}^*\right)^2 \xrightarrow{p} \mathbf{c}'\boldsymbol{\Omega}_\mathbf{2}\mathbf{c}$$

and c_k denote the k-th element of vector $\mathbf{c}$ and ψ_k the k-th of $\mathbf{b}$, then we have the Lyapounov condition for $\delta = 1$

$$\begin{aligned}&\sum_{t=1}^T\frac{1}{s_T^3}\mathscr{E}_*\left|\mathbf{c}'\mathbf{z}_\mathbf{t}^*\right|^3\\ =&\frac{\mathscr{E}_*|\eta_t^{*2}-1|^3}{8s_T^3T^{3/2}}\sum_{t=1}^T\left|\frac{1}{h_t(\widehat{\boldsymbol{\theta}})}\mathbf{c}'\frac{\partial h_t(\widehat{\boldsymbol{\theta}})}{\partial \mathbf{b}}\right|^3\\ \leq&\frac{\mathscr{E}_*|\eta_t^{*2}-1|^3}{8s_T^3\widehat{b_0}^3}\frac{1}{\sqrt{T}}\left(\frac{1}{T}\sum_{t=1}^T\left|\sum_{k=1}^{s+r+1}c_k\frac{\partial h_t(\widehat{\boldsymbol{\theta}})}{\partial \psi_k}\right|^3\right)\\ \leq&\frac{\mathscr{E}_*|\eta_t^{*2}-1|^3 8^{s+r}}{8s_T^3\widehat{b_0}^3}\frac{1}{\sqrt{T}}\sum_{k=1}^{s+r+1}|c_k|^3\left(\frac{1}{T}\sum_{t=1}^T\left|\frac{\partial h_t(\widehat{\boldsymbol{\theta}})}{\partial \psi_k}\right|^3\right)\\ =&o_p(1).\end{aligned}$$

We obtain the last equation from (B1) and Lemma 4.1. The Cramér-Wold theorem and the central limit theorem for triangular arrays applied to the sequence $\mathbf{c}'\mathbf{z}_1^*, ..., \mathbf{c}'\mathbf{z}_T^*$ show

$$\sum_{t=1}^{T} \mathbf{z}_t^* \xrightarrow{d} \mathcal{N}(0, \boldsymbol{\Omega}_2) \quad \textit{in probability.}$$

Secondly, we prove weak consistency of the ARMA part analogously to the proof of the GARCH part. Observe

$$\sqrt{T}(\widehat{\mathbf{a}}^* - \widehat{\mathbf{a}}) = -\widehat{\mathbf{H}_1^*}(\widehat{\boldsymbol{\theta}})^{-1} \frac{1}{\sqrt{T}} \sum_{t=1}^{T} \frac{\partial l_t^*(\widehat{\boldsymbol{\theta}})}{\partial \mathbf{a}}.$$

Note that

$$\frac{\partial \varepsilon_t^*}{\partial \mathbf{a}} = \frac{\partial}{\partial \mathbf{a}} \Big(X_t^* - a_0 - \sum_{i=1}^{p} a_i X_{t-i} - \sum_{j=1}^{q} \alpha_j \varepsilon_{t-j} \Big) = \frac{\partial \varepsilon_t}{\partial \mathbf{a}}$$

and, therefore, $\widehat{\mathbf{H}_1^*}(\widehat{\boldsymbol{\theta}}) = \widehat{\boldsymbol{\Sigma}_1} \xrightarrow{p} \boldsymbol{\Sigma}_1$. Now it suffices to show

$$\frac{1}{\sqrt{T}} \sum_{t=1}^{T} \frac{\partial l_t^*(\widehat{\boldsymbol{\theta}})}{\partial \mathbf{a}} \xrightarrow{d} \mathcal{N}\Big(0, \boldsymbol{\Omega}_1\Big) \quad \textit{in probability.}$$

We observe

$$\begin{aligned}
\mathbf{y}_t^* :=& \frac{1}{\sqrt{T}} \frac{\partial l_t^*(\widehat{\boldsymbol{\theta}})}{\partial \mathbf{a}} \\
=& \frac{1}{\sqrt{T}} \left(\frac{\eta_t^{*2} - 1}{2h_t(\widehat{\boldsymbol{\theta}})} \frac{\partial h_t(\widehat{\boldsymbol{\theta}})}{\partial \mathbf{a}} - \frac{\eta_t^*}{\sqrt{h_t}(\widehat{\boldsymbol{\theta}})} \frac{\partial \varepsilon_t^*(\widehat{\boldsymbol{\theta}})}{\partial \mathbf{a}} \right) \\
=& \frac{1}{\sqrt{T}} \left(\frac{\eta_t^{*2} - 1}{2h_t(\widehat{\boldsymbol{\theta}})} \frac{\partial h_t(\widehat{\boldsymbol{\theta}})}{\partial \mathbf{a}} - \frac{\eta_t^*}{\sqrt{h_t}(\widehat{\boldsymbol{\theta}})} \frac{\partial \varepsilon_t(\widehat{\boldsymbol{\theta}})}{\partial \mathbf{a}} \right),
\end{aligned}$$

t=1, ..., T. For each T the sequence $\mathbf{y}_1^*, ..., \mathbf{y}_T^*$ is independent because $\eta_1^*, ..., \eta_T^*$ is independent. We obtain

$$\mathscr{E}(\mathbf{y}_t^*) = \frac{1}{\sqrt{T}} \left(\frac{\mathscr{E}(\eta_t^{*2}) - 1}{2h_t(\widehat{\boldsymbol{\theta}})} \frac{\partial h_t(\widehat{\boldsymbol{\theta}})}{\partial \mathbf{a}} - \frac{\mathscr{E}(\eta_t^*)}{\sqrt{h_t}(\widehat{\boldsymbol{\theta}})} \frac{\partial \varepsilon_t(\widehat{\boldsymbol{\theta}})}{\partial \mathbf{a}} \right) = 0$$

and

$$
\begin{aligned}
&\sum_{t=1}^{T} \mathscr{E}_*\left(\mathbf{y_t^*}\mathbf{y_t^{*\prime}}\right)\\
=&\left(\mathscr{E}_*(\eta_t^{*4}) - 2\mathscr{E}_*(\eta_t^{*2}) + 1\right)\frac{1}{T}\sum_{t=1}^{T}\left(\frac{1}{4h_t^2(\widehat{\boldsymbol{\theta}})}\frac{\partial h_t(\widehat{\boldsymbol{\theta}})}{\partial \mathbf{a}}\frac{\partial h_t(\widehat{\boldsymbol{\theta}})}{\partial \mathbf{a}'}\right)\\
&-\left(\mathscr{E}_*(\eta_t^{*3}) - \mathscr{E}_*(\eta_t^{*})\right)\frac{1}{T}\sum_{t=1}^{T}\left\{\frac{1}{2(h_t)^{3/2}(\widehat{\boldsymbol{\theta}})}\right.\\
&\left.\left(\frac{\partial h_t(\widehat{\boldsymbol{\theta}})}{\partial \mathbf{a}}\frac{\partial \varepsilon_t(\widehat{\boldsymbol{\theta}})}{\partial \mathbf{a}'} + \frac{\partial \varepsilon_t(\widehat{\boldsymbol{\theta}})}{\partial \mathbf{a}}\frac{\partial h_t(\widehat{\boldsymbol{\theta}})}{\partial \mathbf{a}'}\right)\right\}\\
&+\mathscr{E}_*(\eta_t^{*2})\frac{1}{T}\sum_{t=1}^{T}\left(\frac{1}{h_t(\widehat{\boldsymbol{\theta}})}\frac{\partial \varepsilon_t(\widehat{\boldsymbol{\theta}})}{\partial \mathbf{a}}\frac{\partial \varepsilon_t(\widehat{\boldsymbol{\theta}})}{\partial \mathbf{a}'}\right)\\
&\xrightarrow{p} \boldsymbol{\Omega}_1.
\end{aligned}
$$

Let $\mathbf{c} \in \mathbb{R}^{p+q+1}$,

$$u_T^2 := \sum_{t=1}^{T} \mathscr{E}_*\left(\mathbf{c}'\mathbf{y_t^*}\right)^2 \xrightarrow{p} \mathbf{c}'\boldsymbol{\Omega}_1\mathbf{c}$$

and c_k denote the k-th element of vector $\mathbf{c}$ and ϕ_k the k-th of $\mathbf{a}$, then we have the Lyapounov condition for $\delta = 1$

$$
\begin{aligned}
&\sum_{t=1}^{T} \frac{1}{u_T^3}\mathscr{E}_*\left|\mathbf{c}'\mathbf{y_t^*}\right|^3\\
\leq&\frac{8}{u_T^3 T^{3/2}}\sum_{t=1}^{T}\left(\mathscr{E}_*\left|\mathbf{c}'\frac{\eta_t^{*2}-1}{2h_t(\widehat{\boldsymbol{\theta}})}\frac{\partial h_t(\widehat{\boldsymbol{\theta}})}{\partial \mathbf{a}}\right|^3 + \mathscr{E}_*\left|-\mathbf{c}'\frac{\eta_t^*}{\sqrt{h_t(\widehat{\boldsymbol{\theta}})}}\frac{\partial \varepsilon_t(\widehat{\boldsymbol{\theta}})}{\partial \mathbf{a}}\right|^3\right)\\
\leq&\frac{8}{u_T^3 T^{3/2}}\sum_{t=1}^{T}\left(\mathscr{E}_*|\eta_t^{*2}-1|^3\left|\frac{1}{2h_t(\widehat{\boldsymbol{\theta}})}\mathbf{c}'\frac{\partial h_t(\widehat{\boldsymbol{\theta}})}{\partial \mathbf{a}}\right|^3 + \mathscr{E}_*|\eta_t^*|^3\left|\frac{1}{\sqrt{h_t(\widehat{\boldsymbol{\theta}})}}\mathbf{c}'\frac{\partial \varepsilon_t(\widehat{\boldsymbol{\theta}})}{\partial \mathbf{a}}\right|^3\right)\\
\leq&\frac{\mathscr{E}_*|\eta_t^{*2}-1|^3 8^{p+q}}{u_T^3\widehat{b_0}^3}\frac{1}{\sqrt{T}}\sum_{k=1}^{p+q+1}|c_k|^3\left(\frac{1}{T}\sum_{t=1}^{T}\left|\frac{\partial h_t(\widehat{\boldsymbol{\theta}})}{\partial \phi_k}\right|^3\right)\\
&+\lim_{T\to\infty}\frac{\mathscr{E}_*|\eta_t^*|^3 8^{p+q+1}}{u_T^3\widehat{b_0}^{3/2}}\frac{1}{\sqrt{T}}\sum_{k=1}^{p+q+1}|c_k|^3\left(\frac{1}{T}\sum_{t=1}^{T}\left|\frac{\partial \varepsilon_t(\widehat{\boldsymbol{\theta}})}{\partial \phi_k}\right|^3\right) = o_p(1).
\end{aligned}
$$

We obtain the last equation from (B1) and Lemma 4.1. □

4.3 Wild Bootstrap

Analogously to the previous section, we propose a wild bootstrap technique and investigate its weak consistency. A wild bootstrap method can be applied to the model (4.1) as follows:

(step 1) Obtain the QML estimator $\widehat{\boldsymbol{\theta}}$ and calculate the residuals

$$\widehat{\varepsilon_t} = X_t - \widehat{a_0} - \sum_{i=1}^{p} \widehat{a_i} X_{t-i} - \sum_{j=1}^{q} \widehat{\alpha_j} \widehat{\varepsilon_{t-j}}, \quad t = 1, \ldots, T.$$

(step 2) Compute the estimated heteroscedasticity

$$\widehat{h_t} = \widehat{b_0} + \sum_{j=1}^{s} \widehat{b_j} \widehat{\varepsilon_{t-j}}^2 + \sum_{i=1}^{r} \widehat{\beta_i} \widehat{h_{t-i}}, \quad t = 1, \ldots, T.$$

(step 3) Generate the bootstrap process $X_t^\dagger$ by computing

$$X_t^\dagger = \widehat{a_0} + \sum_{i=1}^{p} \widehat{a_i} X_{t-i} + \sum_{j=1}^{q} \widehat{\alpha_j} \widehat{\varepsilon_{t-j}} + \varepsilon_t^\dagger,$$

$$\varepsilon_t^\dagger = \sqrt{\widehat{h_t}} w_t^\dagger, \quad w_t^\dagger \overset{iid}{\sim} \mathcal{N}(0,1), \quad t = 1, \ldots, T.$$

(step 4) Obtain the bootstrap estimator $\widehat{\mathbf{a}^\dagger}$ by the Newton-Raphson method

$$\widehat{\mathbf{a}^\dagger} = \widehat{\mathbf{a}} - \widehat{\mathbf{H}_1^\dagger}(\widehat{\boldsymbol{\theta}})^{-1} \mathbf{g}_1^\dagger(\widehat{\boldsymbol{\theta}}),$$

where

$$\mathbf{g}_1^\dagger(\widehat{\boldsymbol{\theta}}) = \frac{1}{T} \sum_{t=1}^{T} \frac{\partial l_t^\dagger(\widehat{\boldsymbol{\theta}})}{\partial \mathbf{a}}$$

with

$$\begin{aligned}
\frac{\partial l_t^\dagger(\widehat{\boldsymbol{\theta}})}{\partial \mathbf{a}} &= \frac{1}{2h_t(\widehat{\boldsymbol{\theta}})} \left(\frac{\varepsilon_t^{\dagger 2}}{h_t}(\widehat{\boldsymbol{\theta}}) - 1 \right) \frac{\partial h_t(\widehat{\boldsymbol{\theta}})}{\partial \mathbf{a}} - \frac{\varepsilon_t^\dagger}{h_t}(\widehat{\boldsymbol{\theta}}) \frac{\partial \varepsilon_t^\dagger(\widehat{\boldsymbol{\theta}})}{\partial \mathbf{a}} \\
&= \frac{\eta_t^{\dagger 2} - 1}{2h_t(\widehat{\boldsymbol{\theta}})} \frac{\partial h_t(\widehat{\boldsymbol{\theta}})}{\partial \mathbf{a}} - \frac{\eta_t^\dagger}{\sqrt{h_t}(\widehat{\boldsymbol{\theta}})} \frac{\partial \varepsilon_t^\dagger(\widehat{\boldsymbol{\theta}})}{\partial \mathbf{a}}
\end{aligned}$$

and

$$\widehat{\mathbf{H}_1^\dagger}(\widehat{\boldsymbol{\theta}}) = \frac{1}{T} \sum_{t=1}^{T} \left(\frac{1}{2h_t^2(\widehat{\boldsymbol{\theta}})} \frac{\partial h_t(\widehat{\boldsymbol{\theta}})}{\partial \mathbf{a}} \frac{\partial h_t(\widehat{\boldsymbol{\theta}})}{\partial \mathbf{a}'} + \frac{1}{h_t(\widehat{\boldsymbol{\theta}})} \frac{\partial \varepsilon_t^\dagger(\widehat{\boldsymbol{\theta}})}{\partial \mathbf{a}} \frac{\partial \varepsilon_t^\dagger(\widehat{\boldsymbol{\theta}})}{\partial \mathbf{a}'} \right).$$

(step 5) Analogously, calculate the bootstrap estimator $\widehat{\mathbf{b}^\dagger}$

$$\widehat{\mathbf{b}^\dagger} = \widehat{\mathbf{b}} - \widehat{\mathbf{H}_2^\dagger}(\widehat{\boldsymbol{\theta}})^{-1}\mathbf{g}_2^\dagger(\widehat{\boldsymbol{\theta}}),$$

where

$$\mathbf{g}_2^\dagger(\widehat{\boldsymbol{\theta}}) = \frac{1}{T}\sum_{t=1}^{T}\frac{\partial l_t^\dagger(\widehat{\boldsymbol{\theta}})}{\partial \mathbf{b}}$$

with

$$\begin{aligned}\frac{\partial l_t^\dagger(\widehat{\boldsymbol{\theta}})}{\partial \mathbf{b}} &= \frac{1}{2h_t(\widehat{\boldsymbol{\theta}})}\left(\frac{\varepsilon_t^{\dagger 2}}{h_t}(\widehat{\boldsymbol{\theta}}) - 1\right)\frac{\partial h_t(\widehat{\boldsymbol{\theta}})}{\partial \mathbf{b}} \\ &= \frac{\eta_t^{\dagger 2} - 1}{2h_t(\widehat{\boldsymbol{\theta}})}\frac{\partial h_t(\widehat{\boldsymbol{\theta}})}{\partial \mathbf{b}}\end{aligned}$$

and

$$\widehat{\mathbf{H}_2^\dagger}(\widehat{\boldsymbol{\theta}}) = \frac{1}{T}\sum_{t=1}^{T}\left(\frac{1}{2h_t^2(\widehat{\boldsymbol{\theta}})}\frac{\partial h_t(\widehat{\boldsymbol{\theta}})}{\partial \mathbf{b}}\frac{\partial h_t(\widehat{\boldsymbol{\theta}})}{\partial \mathbf{b}'}\right).$$

Theorem 4.4

Suppose that X_t is generated by the model (4.1) satisfying assumptions (A1)-(A5). Then the wild bootstrap of section 4.3 has the following asymptotic property:

$$\begin{aligned}&\sqrt{T}(\widehat{\mathbf{a}^\dagger} - \widehat{\mathbf{a}}) \xrightarrow{d} \mathcal{N}\left(0, \boldsymbol{\Sigma}_1^{-1}\left(\boldsymbol{\Omega}_1 + \boldsymbol{\Omega}_3\right)\boldsymbol{\Sigma}_1^{-1}\right) \quad \textit{in probability,} \\ &\sqrt{T}(\widehat{\mathbf{b}^\dagger} - \widehat{\mathbf{b}}) \xrightarrow{d} \mathcal{N}\left(0, \boldsymbol{\Sigma}_2^{-1}\tau\boldsymbol{\Omega}_2\boldsymbol{\Sigma}_2^{-1}\right) \quad \textit{in probability,}\end{aligned}$$

where

$$\boldsymbol{\Omega}_3 = \frac{3-\kappa}{2}\mathscr{E}\left(\frac{1}{2h_t^2(\boldsymbol{\theta}_0)}\frac{\partial h_t(\boldsymbol{\theta}_0)}{\partial \mathbf{a}}\frac{\partial h_t(\boldsymbol{\theta}_0)}{\partial \mathbf{a}'}\right) \quad \textit{and} \quad \tau = \frac{2}{\kappa - 1}.$$

Proof. The proof is the mirror image of Theorem 4.3. Firstly, we observe

$$\sqrt{T}(\widehat{\mathbf{b}^\dagger} - \widehat{\mathbf{b}}) = -\widehat{\mathbf{H}_2^\dagger}(\widehat{\boldsymbol{\theta}})^{-1}\frac{1}{\sqrt{T}}\sum_{t=1}^{T}\frac{\partial l_t^\dagger(\widehat{\boldsymbol{\theta}})}{\partial \mathbf{b}}$$

and prove

$$\frac{1}{\sqrt{T}}\sum_{t=1}^{T}\frac{\partial l_t^\dagger(\widehat{\boldsymbol{\theta}})}{\partial \mathbf{b}} \xrightarrow{d} \mathcal{N}\left(0, \tau\boldsymbol{\Omega}_2\right) \quad \textit{in probability.}$$

Let

$$\begin{aligned}\mathbf{z}_\mathbf{t}^\dagger :=& \frac{1}{\sqrt{T}}\frac{\partial l_t^\dagger(\widehat{\boldsymbol{\theta}})}{\partial \mathbf{b}}\\ =& \frac{1}{\sqrt{T}}\frac{w_t^{\dagger 2}-1}{2h_t(\widehat{\boldsymbol{\theta}})}\frac{\partial h_t(\widehat{\boldsymbol{\theta}})}{\partial \mathbf{b}}, \qquad t=1,...,T.\end{aligned}$$

For each T the sequence $\mathbf{z}_\mathbf{1}^\dagger,...,\mathbf{z}_\mathbf{T}^\dagger$ is independent because $w_1^\dagger,...,w_T^\dagger$ is independent. We obtain $\mathscr{E}_\dagger(\mathbf{z}_\mathbf{t}^\dagger)=0$ and

$$\begin{aligned}&\sum_{t=1}^T \mathscr{E}_\dagger\left(\mathbf{z}_\mathbf{t}^\dagger \mathbf{z}_\mathbf{t}^{\dagger\prime}\right)\\ =&\left(\mathscr{E}_\dagger(w_t^{\dagger 4})-2\mathscr{E}_\dagger(w_t^{\dagger 2})+1\right)\frac{1}{T}\sum_{t=1}^T\left(\frac{1}{4h_t^2(\widehat{\boldsymbol{\theta}})}\frac{\partial h_t(\widehat{\boldsymbol{\theta}})}{\partial \mathbf{b}}\frac{\partial h_t(\widehat{\boldsymbol{\theta}})}{\partial \mathbf{b}'}\right)\\ =&\frac{1}{T}\sum_{t=1}^T\left(\frac{1}{2h_t^2(\widehat{\boldsymbol{\theta}})}\frac{\partial h_t(\widehat{\boldsymbol{\theta}})}{\partial \mathbf{b}}\frac{\partial h_t(\widehat{\boldsymbol{\theta}})}{\partial \mathbf{b}'}\right)\\ \xrightarrow{p}&\tau\boldsymbol{\Omega}_\mathbf{2}.\end{aligned}$$

Let $\mathbf{c}\in\mathbb{R}^{s+r+1}$ and

$$s_T^2 := \sum_{t=1}^T \mathscr{E}_\dagger\left(\mathbf{c}'\mathbf{z}_\mathbf{t}^\dagger\right)^2 \xrightarrow{p} \mathbf{c}'\tau\boldsymbol{\Omega}_\mathbf{2}\mathbf{c},$$

then we have the Lyapounov condition for $\delta=1$

$$\begin{aligned}&\sum_{t=1}^T\frac{1}{s_T^3}\mathscr{E}_\dagger\left|\mathbf{c}'\mathbf{z}_\mathbf{t}^\dagger\right|^3\\ =&\frac{\mathscr{E}_\dagger|w_t^{\dagger 2}-1|^3}{8s_T^3T^{3/2}}\sum_{t=1}^T\left|\frac{1}{h_t(\widehat{\boldsymbol{\theta}})}\mathbf{c}'\frac{\partial h_t(\widehat{\boldsymbol{\theta}})}{\partial \mathbf{b}}\right|^3\\ =&o_p(1).\end{aligned}$$

Secondly, we observe

$$\sqrt{T}(\widehat{\mathbf{a}}^\dagger-\widehat{\mathbf{a}}) = -\widehat{\mathbf{H}_\mathbf{1}^\dagger}(\widehat{\boldsymbol{\theta}})^{-1}\frac{1}{\sqrt{T}}\sum_{t=1}^T\frac{\partial l_t^\dagger(\widehat{\boldsymbol{\theta}})}{\partial \mathbf{a}}$$

and prove

$$\frac{1}{\sqrt{T}}\sum_{t=1}^T\frac{\partial l_t^\dagger(\widehat{\boldsymbol{\theta}})}{\partial \mathbf{a}} \xrightarrow{d} \mathscr{N}\left(0,\boldsymbol{\Omega}_\mathbf{1}+\boldsymbol{\Omega}_\mathbf{3}\right) \qquad \textit{in probability.}$$

Let

$$\mathbf{y_t^\dagger} := \frac{1}{\sqrt{T}} \frac{\partial l_t^\dagger(\widehat{\boldsymbol{\theta}})}{\partial \mathbf{a}} = \frac{1}{\sqrt{T}} \left(\frac{w_t^{\dagger 2} - 1}{2h_t(\widehat{\boldsymbol{\theta}})} \frac{\partial h_t(\widehat{\boldsymbol{\theta}})}{\partial \mathbf{a}} - \frac{w_t^\dagger}{\sqrt{h_t}(\widehat{\boldsymbol{\theta}})} \frac{\partial \varepsilon_t(\widehat{\boldsymbol{\theta}})}{\partial \mathbf{a}} \right), \quad t = 1, ..., T.$$

For each T the sequence $\mathbf{y_1^\dagger}, ..., \mathbf{y_T^\dagger}$ is independent because $w_1^\dagger, ..., w_T^\dagger$ is independent. We obtain $\mathscr{E}_\dagger(\mathbf{y_t^\dagger}) = 0$ and

$$\begin{aligned}
&\sum_{t=1}^{T} \mathscr{E}_\dagger\left(\mathbf{y_t^\dagger} \mathbf{y_t^{\dagger\prime}}\right) \\
=&\left(\mathscr{E}_\dagger(w_t^{\dagger 4}) - 2\mathscr{E}_\dagger(w_t^{\dagger 2}) + 1\right) \frac{1}{T} \sum_{t=1}^{T} \left(\frac{1}{4h_t^2(\widehat{\boldsymbol{\theta}})} \frac{\partial h_t(\widehat{\boldsymbol{\theta}})}{\partial \mathbf{a}} \frac{\partial h_t(\widehat{\boldsymbol{\theta}})}{\partial \mathbf{a}'} \right) \\
&- \left(\mathscr{E}_\dagger(w_t^{\dagger 3}) - \mathscr{E}_\dagger(w_t^\dagger)\right) \frac{1}{T} \sum_{t=1}^{T} \left\{ \frac{1}{2(h_t)^{3/2}(\widehat{\boldsymbol{\theta}})} \right. \\
&\left. \left(\frac{\partial h_t(\widehat{\boldsymbol{\theta}})}{\partial \mathbf{a}} \frac{\partial \varepsilon_t(\widehat{\boldsymbol{\theta}})}{\partial \mathbf{a}'} + \frac{\partial \varepsilon_t(\widehat{\boldsymbol{\theta}})}{\partial \mathbf{a}} \frac{\partial h_t(\widehat{\boldsymbol{\theta}})}{\partial \mathbf{a}'} \right) \right\} \\
&+ \mathscr{E}_\dagger(w_t^{\dagger 2}) \frac{1}{T} \sum_{t=1}^{T} \left(\frac{1}{h_t(\widehat{\boldsymbol{\theta}})} \frac{\partial \varepsilon_t(\widehat{\boldsymbol{\theta}})}{\partial \mathbf{a}} \frac{\partial \varepsilon_t(\widehat{\boldsymbol{\theta}})}{\partial \mathbf{a}'} \right) \\
=&\frac{1}{T} \sum_{t=1}^{T} \left(\frac{1}{2h_t^2(\widehat{\boldsymbol{\theta}})} \frac{\partial h_t(\widehat{\boldsymbol{\theta}})}{\partial \mathbf{a}} \frac{\partial h_t(\widehat{\boldsymbol{\theta}})}{\partial \mathbf{a}'} \right) + \frac{1}{T} \sum_{t=1}^{T} \left(\frac{1}{h_t(\widehat{\boldsymbol{\theta}})} \frac{\partial \varepsilon_t(\widehat{\boldsymbol{\theta}})}{\partial \mathbf{a}} \frac{\partial \varepsilon_t(\widehat{\boldsymbol{\theta}})}{\partial \mathbf{a}'} \right) \\
&\xrightarrow{p} \boldsymbol{\Omega}_1 + \boldsymbol{\Omega}_3.
\end{aligned}$$

Let $\mathbf{c} \in \mathbb{R}^{p+q+1}$ and

$$u_T^2 := \sum_{t=1}^{T} \mathscr{E}_\dagger\left(\mathbf{c}'\mathbf{y_t^\dagger}\right)^2 \xrightarrow{p} \mathbf{c}'\left(\boldsymbol{\Omega}_1 + \boldsymbol{\Omega}_3\right)\mathbf{c},$$

then we have the Lyapounov condition for $\delta = 1$

$$
\begin{aligned}
&\sum_{t=1}^{T} \frac{1}{u_T^3} \mathscr{E}_\dagger \left| \mathbf{c}' \mathbf{y}_t^\dagger \right|^3 \\
\leq & \frac{8}{u_T^3 T^{3/2}} \sum_{t=1}^{T} \left(\mathscr{E}_\dagger \left| \mathbf{c}' \frac{w_t^{\dagger 2} - 1}{2h_t(\widehat{\boldsymbol{\theta}})} \frac{\partial h_t(\widehat{\boldsymbol{\theta}})}{\partial \mathbf{a}} \right|^3 + \mathscr{E}_\dagger \left| -\mathbf{c}' \frac{w_t^\dagger}{\sqrt{h_t}(\widehat{\boldsymbol{\theta}})} \frac{\partial \varepsilon_t(\widehat{\boldsymbol{\theta}})}{\partial \mathbf{a}} \right|^3 \right) \\
\leq & \frac{8}{u_T^3 T^{3/2}} \sum_{t=1}^{T} \left(\mathscr{E}_\dagger |w_t^{\dagger 2} - 1|^3 \left| \frac{1}{2h_t(\widehat{\boldsymbol{\theta}})} \mathbf{c}' \frac{\partial h_t(\widehat{\boldsymbol{\theta}})}{\partial \mathbf{a}} \right|^3 + \mathscr{E}_\dagger |w_t^\dagger|^3 \left| \frac{1}{\sqrt{h_t}(\widehat{\boldsymbol{\theta}})} \mathbf{c}' \frac{\partial \varepsilon_t(\widehat{\boldsymbol{\theta}})}{\partial \mathbf{a}} \right|^3 \right) \\
= & o_p(1).
\end{aligned}
$$

□

Remark 4.3

Theorem 4.4 implies that the wild bootstrap of section 4.3 is weakly consistent only if $\kappa = 3$. In other words, if $\kappa \neq 3$, it is necessary to construct a wild bootstrap in a different way so that $\mathscr{E}_\dagger(w_t^\dagger) = 0$, $\mathscr{E}_\dagger(w_t^{\dagger 2}) = 1$, $\mathscr{E}_\dagger(w_t^{\dagger 3}) \xrightarrow{p} 0$ and $\mathscr{E}_\dagger(w_t^{\dagger 4}) \xrightarrow{p} \kappa$; however, it is not always easy to obtain such $w_t^\dagger$. In the next section we will see the result of a false application, in which the wild bootstrap with standard normal distribution is applied though $\kappa = 9$.

4.4 Simulations

In this section we demonstrate the large sample properties of the residual bootstrap of section 4.2 and the wild bootstrap of section 4.3. Let us consider the following ARMA(1,1)-GARCH(1,1) model:

$$
\begin{aligned}
X_t &= a_0 + a_1 X_{t-1} + \alpha_1 \varepsilon_{t-1} + \varepsilon_t, \\
\varepsilon_t &= \sqrt{h_t} \eta_t, \\
h_t &= b_0 + b_1 \varepsilon_{t-1}^2 + \beta_1 h_{t-1}, \qquad t = 1, ..., 10,000.
\end{aligned}
$$

The simulation procedure is as follows.

(step 1) Simulate the bias $\eta_t \overset{iid}{\sim} \sqrt{\frac{3}{5}}\, t_5$ for $t = 1, ..., 10,000$.

(step 2) Let $X_0 = \mathscr{E}(X_t) = \frac{a_0}{1-a_1}$, $h_0 = \varepsilon_0^2 = \mathscr{E}(\varepsilon_t^2) = \frac{b_0}{1-b_1-\beta}$, $a_0 = 0.141$, $a_1 = 0.433$, $\alpha_1 = -0.162$, $b_0 = 0.007$, $b_1 = 0.135$ and $\beta_1 = 0.829$ and calculate

$$
X_t = a_0 + a_1 X_{t-1} + \alpha_1 \varepsilon_{t-1} + \left(b_0 + b_1 \varepsilon_{t-1}^2 + \beta_1 h_{t-1} \right)^{1/2} \eta_t
$$

for $t = 1, ..., 10,000$.

(step 3) From the data set $\{X_1, X_2, ..., X_{10,000}\}$ obtain the QML estimators $\widehat{a_0}$, $\widehat{a_1}$, $\widehat{\alpha_1}$, $\widehat{b_0}$, $\widehat{b_1}$ and $\widehat{\beta_1}$.

(step 4) Repeat (step 1)-(step 3) 2,000 times.

(step 5) Choose one data set $\{X_1, X_2, ..., X_{10,000}\}$ and its estimators $\{\widehat{a_0}, \widehat{a_1}, \widehat{\alpha_1}, \widehat{b_0}, \widehat{b_1}, \widehat{\beta_1}\}$. Adopt the residual bootstrap of section 4.2 and the wild bootstrap of section 4.3 based on the chosen data set and estimators. Both bootstrap methods are repeated 2,000 times.

(step 6) Compare the simulated density of the QML estimators ($\sqrt{T}(\widehat{a_0} - a_0)$ etc., thin lines) with the residual and the wild bootstrap approximations ($\sqrt{T}(\widehat{a_0^*} - \widehat{a_0})$, $\sqrt{T}(\widehat{a_0^\dagger} - \widehat{a_0})$ etc., bold lines).

Figure 4.1 indicates that the residual bootstrap is weakly consistent. On the other hand, Figure 4.2 shows the problem of the wild bootstrap, especially for the GARCH part, which corresponds to the theoretical result in Theorem 4.4 with $\tau = \frac{1}{4}$. Note that in this simulation $\mathbf{\Omega_3}$ is much smaller than $\mathbf{\Omega_1}$, i.e. $\mathbf{\Omega_1} \approx \mathbf{\Omega_1} + \mathbf{\Omega_3}$. Thus, we do not see the problem of the ARMA part, which theoretically exists, in Figure 4.2.

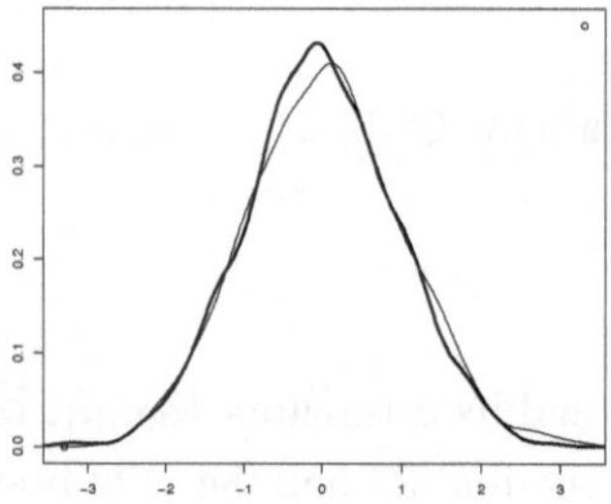

(a) $\sqrt{T}(\widehat{a_0}-a_0)$ and $\sqrt{T}(\widehat{a_0^*}-\widehat{a_0})$

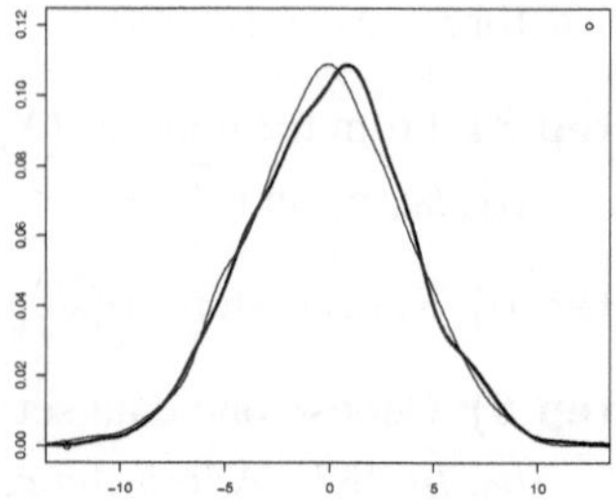

(b) $\sqrt{T}(\widehat{a_1}-a_1)$ and $\sqrt{T}(\widehat{a_1^*}-\widehat{a_1})$

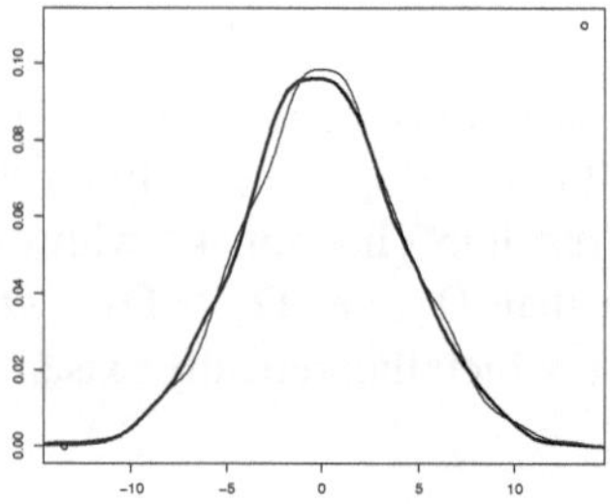

(c) $\sqrt{T}(\widehat{\alpha_1}-\alpha_1)$ and $\sqrt{T}(\widehat{\alpha_1^*}-\widehat{\alpha_1})$

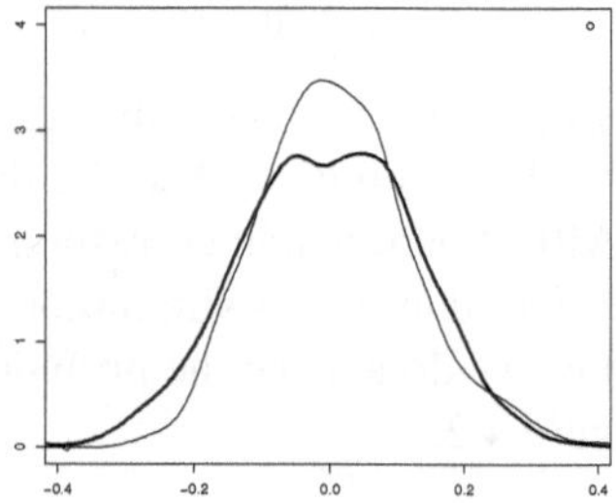

(d) $\sqrt{T}(\widehat{b_0}-b_0)$ and $\sqrt{T}(\widehat{b_0^*}-\widehat{b_0})$

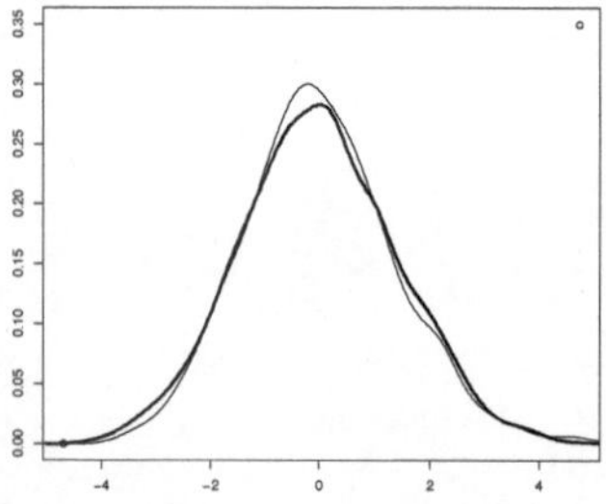

(e) $\sqrt{T}(\widehat{b_1}-b_1)$ and $\sqrt{T}(\widehat{b_1^*}-\widehat{b_1})$

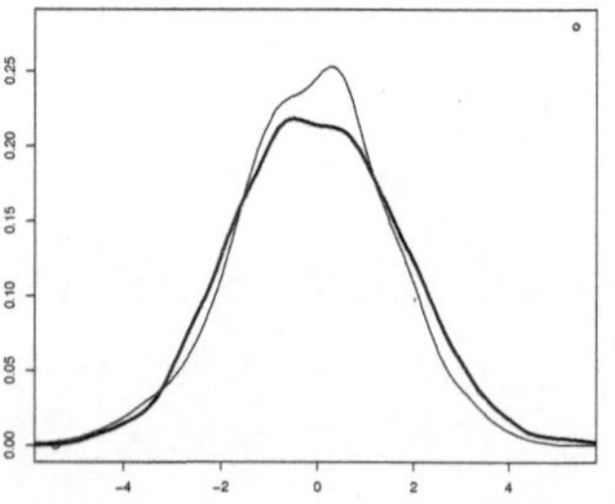

(f) $\sqrt{T}(\widehat{\beta_1}-\beta_1)$ and $\sqrt{T}(\widehat{\beta_1^*}-\widehat{\beta_1})$

Figure 4.1: Residual Bootstrap

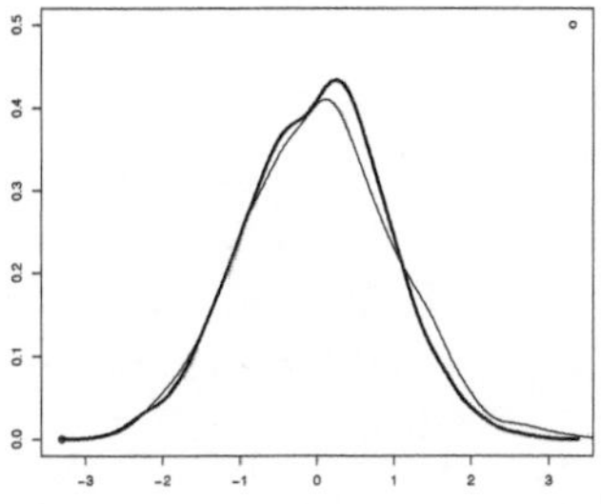

(a) $\sqrt{T}(\widehat{a_0} - a_0)$ and $\sqrt{T}(\widehat{a_0^\dagger} - \widehat{a_0})$

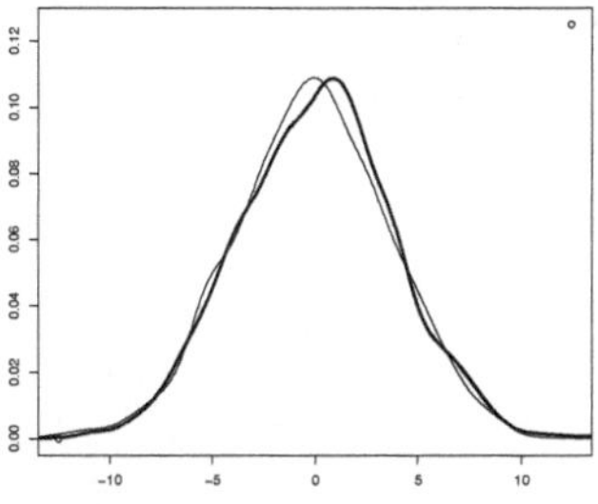

(b) $\sqrt{T}(\widehat{a_1} - a_1)$ and $\sqrt{T}(\widehat{a_1^\dagger} - \widehat{a_1})$

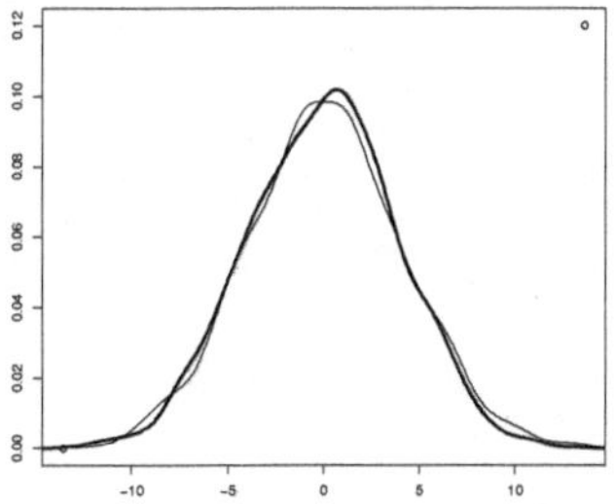

(c) $\sqrt{T}(\widehat{\alpha_1} - \alpha_1)$ and $\sqrt{T}(\widehat{\alpha_1^\dagger} - \widehat{\alpha_1})$

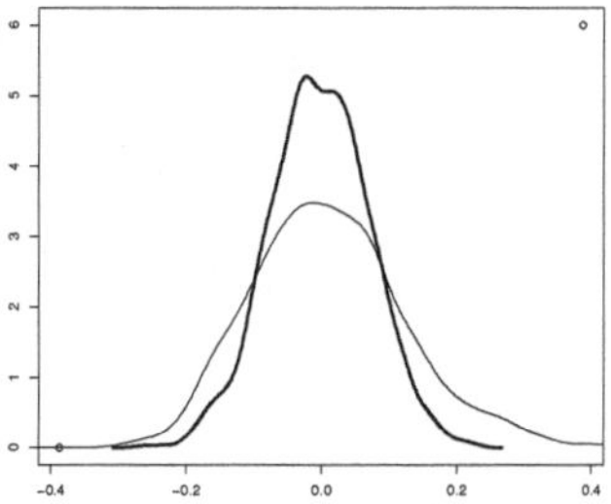

(d) $\sqrt{T}(\widehat{b_0} - b_0)$ and $\sqrt{T}(\widehat{b_0^\dagger} - \widehat{b_0})$

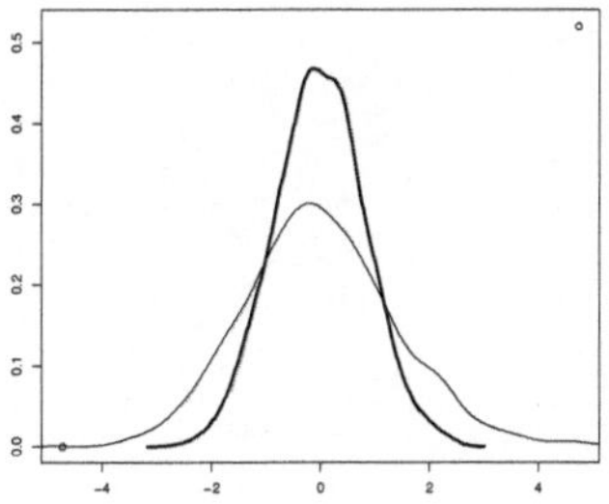

(e) $\sqrt{T}(\widehat{b_1} - b_1)$ and $\sqrt{T}(\widehat{b_1^\dagger} - \widehat{b_1})$

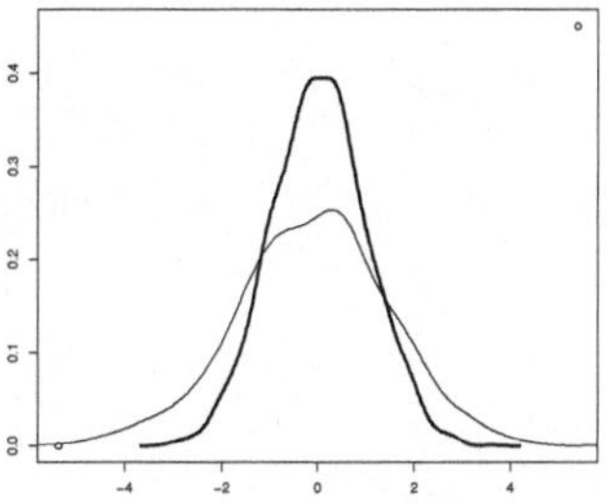

(f) $\sqrt{T}(\widehat{\beta_1} - \beta_1)$ and $\sqrt{T}(\widehat{\beta_1^\dagger} - \widehat{\beta_1})$

Figure 4.2: Wild Bootstrap

5 Semiparametric AR(p)-ARCH(1) Models

So far we have analysed parametric ARCH and GARCH models. In this chapter the results are extended to semiparametric models, in which the ARCH part is nonparametric. In the first section we introduce the semiparametric AR(p)-ARCH(1) model and show the asymptotic properties of the estimators. Then, as in preceding chapters, possible applications of the residual and the wild bootstrap are proposed and their weak consistency proved. The theoretical results are confirmed by simulations in the last section.

5.1 Estimation Theory

Franke, Kreiß and Mammen (2002) investigated asymptotic properties of the Nadaraya-Watson (NW) estimator for the first order nonparametric autoregressive conditional heteroscedasticity (NARCH(1)) model of the form

$$X_t = m(X_{t-1}) + \sigma(X_{t-1})\eta_t,$$

where $\{\eta_t\}$ is a sequence of i.i.d. random variables with mean 0 and variance 1, and m and σ are unknown smooth functions. They applied the residual, the wild and the autoregression bootstrap techniques to the model and proved their weak consistency. Palkowski (2005) extended the analysis to Fan and Yao's (1998) estimator and applied the wild bootstrap technique to the model.

It is important to note that the NARCH(1) model is fundamentally different from Engle's ARCH(1) model

$$\begin{aligned} X_t &= a_1 X_{t-1} + \sqrt{h_t}\eta_t, \\ h_t &= b_0 + b_1 \varepsilon_{t-1}^2, \end{aligned}$$

where $\{\eta_t\}$ is a sequence of i.i.d. random variables with mean 0 and variance 1, and a_1, b_0 and b_1 are unknown parameters. While the NARCH model assumes that variance of the disturbance at time t is a function of the observable random variable at time *t-1* $\left(\sigma^2(X_{t-1})\right)$, Engle's ARCH model assumes that this is a function of the square of the disturbance at time *t-1* $\left(b_0 + b_1\varepsilon_{t-1}^2\right)$. Financial time series is characterised by volatility clustering, that is, "large changes tend to be followed by large changes, of either sign, and small changes tend to be followed by small changes".[1] Volatility clustering suggests a time series in which successive disturbances are serially dependent[2] and, therefore, we propose the new semiparametric AR-ARCH model, in which the AR part is parametric, the ARCH part is nonparametric and variance of the disturbance at time t is a function of the disturbance at time *t-1* $\left(\sigma^2(\varepsilon_{t-1})\right)$. Note that the new semiparametric AR-ARCH model is more general than Engle's ARCH model in that the former analyses the relationship between ε_t and ε_{t-1} nonparametrically while the latter between ε_t^2 and ε_{t-1}^2 parametrically.[3]

5.1.1 Model and Assumptions

Assume that $\{X_t,\ t = 0,...,T\}$ are generated by the p-th order parametric autoregressive model with the NARCH(1) errors:

$$\begin{aligned} X_t &= a_0 + \sum_{i=1}^{p} a_i X_{t-i} + \varepsilon_t, \\ \varepsilon_t &= \sigma(\varepsilon_{t-1})\eta_t, \end{aligned} \tag{5.1}$$

where $\{\eta_t\}$ is a sequence of i.i.d. random variables such that $\mathscr{E}(\eta_t) = 0$, $\mathscr{E}(\eta_t^2) = 1$, $\mathscr{E}(\eta_t^4) =: \kappa < \infty$ and σ is an unknown smooth function. For simplicity we assume here the presample values $\{X_t,\ t = -p,\ ...,\ -1\}$.

The process $\{\varepsilon_t\}$ is assumed to be stationary and geometrically ergodic. Let π denote the unique stationary distribution. Stationarity and geometric ergodicity of $\{\varepsilon_t\}$ follow from the following two conditions (cf. Franke, Kreiß and Mammen (2002), section 2.1):

(A1) The distribution of the i.i.d. innovations η_t possesses a Lebesgue density p_η, which satisfies

$$\inf_{e \in C} p_\eta(e) > 0$$

for all compact sets C.

[1] Mandelbrot (1963).

[2] See, e.g., Chang (2006).

[3] Note that this model differs from the semiparametric ARCH models of Engle and González-Rivera (1991) in that the former model is estimated with the NW method based on the estimated residuals of the AR part while the latter model adopts the discrete maximum penalised likelihood estimation technique of Tapia and Thompson (1978).

(A2) σ and σ^{-1} are bounded on compact sets and

$$\limsup_{|e|\to\infty} \frac{\mathscr{E}|\sigma(e)\eta_1|}{|e|} < 1,$$

that is,

$$\limsup_{|e|\to\infty} \frac{\sigma^2(e)}{e^2} < 1.$$

(A1) and (A2) ensure that the stationary distribution π possesses a strictly positive Lebesgue density, which we denote p. From (5.1) we obtain

$$p(e) = \int_{\mathbb{R}} \frac{1}{\sigma(u)} p_\eta \left(\frac{e}{\sigma(u)}\right) p(u)du.$$

We further assume that

(B1) σ is twice continuously differentiable and σ, σ' and σ'' are bounded. There exists $\sigma_0 > 0$ such that $\sigma(e) \geq \sigma_0$ for all $e \in \mathbb{R}$.

(B2) $\mathscr{E}(\eta_1^6) < \infty$. p_η is twice continuously differentiable. p_η, p'_η and p''_η are bounded and

$$\sup_{e\in\mathbb{R}} |ep'_\eta(e)| < \infty.$$

(B3) $g, h \to 0$, $Th^4 \to \infty$, $Th^5 \to B^2 \geq 0$ and $g \sim T^{-\alpha}$ with $0 < \alpha \leq \frac{1}{9}$ for $T \to \infty$.

(B4) The kernel function K has compact support [-1, 1]. K is symmetric, non-negative and three times continuously differentiable with $K(1) = K'(1) = 0$ and $\int K(v)dv = 1$.

5.1.2 NW Estimation

In the semiparametric AR(p)-ARCH(1) model it is necessary to estimate the parametric AR part firstly. The nonparametric ARCH part is then estimated based on the residuals of the AR part. Because we have already analysed the parametric AR(p) model with heteroscedastic errors in chapter 3, we concentrate here on the nonparametric ARCH part. For simplicity of the discussion we adopt the OLS estimation for the AR(p) part and make use of the results of chapter 3.

5.1.2.1 The imaginary case where ε_t are known

Firstly, we analyse the imaginary case where $\{\varepsilon_t,\ t = 1,\ ...,\ T\}$ are known. Let $m_t := \frac{\eta_t^2 - 1}{\sqrt{\kappa - 1}}$, then $\mathscr{E}(m_t) = 0$, $\mathscr{E}(m_t^2) = 1$ and

$$\varepsilon_t^2 = \sigma^2(\varepsilon_{t-1}) + \sqrt{\kappa - 1}\sigma^2(\varepsilon_{t-1})m_t, \quad t = 1, ..., T. \tag{5.2}$$

The NW estimator of σ^2 is

$$\widehat{\sigma_h^2}(e) = \frac{T^{-1}\sum_{i=1}^T K_h(e - \varepsilon_{i-1})\varepsilon_i^2}{T^{-1}\sum_{j=1}^T K_h(e - \varepsilon_{j-1})},$$

where $K_h(\bullet) = \frac{1}{h}K(\bullet/h)$, $K(\bullet)$ is a kernel function and h denotes the bandwidth. Now we sketch the result of Franke, Kreiß and Mammen (2002) observing the decomposition

$$\begin{aligned}
&\sqrt{Th}\left(\widehat{\sigma_h^2}(e) - \sigma^2(e)\right) \\
&= \frac{\sqrt{\frac{h}{T}}\sum_{i=1}^T K_h(e - \varepsilon_{i-1})\sigma^2(\varepsilon_{i-1})(\eta_i^2 - 1)}{\frac{1}{T}\sum_{j=1}^T K_h(e - \varepsilon_{j-1})} \\
&\quad + \frac{\sqrt{\frac{h}{T}}\sum_{i=1}^T K_h(e - \varepsilon_{i-1})\left(\sigma^2(\varepsilon_{i-1}) - \sigma^2(e)\right)}{\frac{1}{T}\sum_{j=1}^T K_h(e - \varepsilon_{j-1})}
\end{aligned}$$

and analysing asymptotic properties of each part separately.

Lemma 5.1

Suppose that ε_t is generated by the model (5.2) satisfying assumptions (A1)-(A2) and (B1)-(B4). Then for all $e \in \mathbb{R}$

$$\sqrt{\frac{h}{T}}\sum_{t=1}^T K_h(e - \varepsilon_{t-1})\sigma^2(\varepsilon_{t-1})(\eta_t^2 - 1) \xrightarrow{d} \mathscr{N}\left(0, \tau^2(e)\right),$$

where $\tau^2(e) = (\kappa - 1)\sigma^4(e)p(e)\int K^2(v)dv$.

Proof. Franke, Kreiß and Mammen (1997), Lemma 6.2 (i). See also Lemma 2.1 in Palkowski (2005). □

Lemma 5.2

Suppose that ε_t is generated by the model (5.2) satisfying assumptions (A1)-(A2) and (B1)-(B4). Then for all $e \in \mathbb{R}$

$$\frac{1}{T}\sum_{t=1}^T K_h(e - \varepsilon_{t-1}) \xrightarrow{p} p(e).$$

Proof. Franke, Kreiß and Mammen (1997), Lemma 6.3 (i). See also Lemma 2.3 in Palkowski (2005). □

Lemma 5.3

Suppose that ε_t is generated by the model (5.2) satisfying assumptions (A1)-(A2) and (B1)-(B4). Then for all $e \in \mathbb{R}$

$$\sqrt{\frac{h}{T}} \sum_{t=1}^{T} K_h(e-\varepsilon_{t-1})\Big(\sigma^2(\varepsilon_{t-1})-\sigma^2(e)\Big) \xrightarrow{p} b(e),$$

where $b(e) = B\int v^2 K(v)dv\Big(p'(e)(\sigma^2)'(e)+\frac{1}{2}p(e)(\sigma^2)''(e)\Big)$.

Proof. Franke, Kreiß and Mammen (1997), Lemma 6.4 (i). See also Lemma 2.2 in Palkowski (2005). □

Theorem 5.1

Suppose that ε_t is generated by the model (5.2) satisfying assumptions (A1)-(A2) and (B1)-(B4). Then for all $e \in \mathbb{R}$

$$\sqrt{Th}\Big(\widehat{\sigma_h^2}(e)-\sigma^2(e)\Big) \xrightarrow{d} \mathcal{N}\left(\frac{b(e)}{p(e)}, \frac{\tau^2(e)}{p^2(e)}\right).$$

Proof. Lemma 5.1-5.3 together with the Slutsky theorem yield the desired result. □

5.1.2.2 The standard case where ε_t are unknown

Secondly, we analyse the standard case where $\{\varepsilon_t,\ t=1,\ ...,\ T\}$ are unknown. In this case we adopt the two-step estimation, where the residuals of the AR part

$$\widehat{\varepsilon_t} = X_t - \widehat{a_0} - \sum_{i=1}^{p}\widehat{a_i}X_{t-i}, \qquad t = 0,\dots,T$$

are used to estimate the ARCH part. The NW estimator of σ^2 is then

$$\widetilde{\sigma_h^2}(e) = \frac{T^{-1}\sum_{i=1}^{T}K_h(e-\widehat{\varepsilon_{i-1}})\widehat{\varepsilon_i}^2}{T^{-1}\sum_{j=1}^{T}K_h(e-\widehat{\varepsilon_{j-1}})}.$$

We observe

$$\sqrt{Th}\Big(\widetilde{\sigma_h^2}(e)-\sigma^2(e)\Big) = \frac{\sqrt{\frac{h}{T}}\sum_{i=1}^{T}K_h(e-\widehat{\varepsilon_{i-1}})\Big(\widehat{\varepsilon_i}^2-\sigma^2(e)\Big)}{\frac{1}{T}\sum_{j=1}^{T}K_h(e-\widehat{\varepsilon_{j-1}})}$$

and analyse the asymptotic properties of each part separately. In fact, the estimator $\widetilde{\sigma_h^2}(e)$ is asymptotically as efficient as $\widehat{\sigma_h^2}(e)$ in §5.1.2.1, as we will see in Theorem 5.2.

Remark 5.1

Analogously to chapter 3, we denote

$$\varepsilon_t - \widehat{\varepsilon_t} = \mathbf{x_t}'(\widehat{\mathbf{a}} - \mathbf{a}) = \sum_{k=0}^{p} x_{t,k}(\widehat{a_k} - a_k),$$

where

$$\mathbf{x_t} = (1\ X_{t-1}\ X_{t-2}\ \dots\ X_{t-p})' = (x_{t,0}\ x_{t,1}\ x_{t,2}\ \dots\ x_{t,p})',$$
$$\mathbf{a} = (a_0\ a_1\ \dots\ a_p)' \quad \text{and} \quad \widehat{\mathbf{a}} = (\widehat{a_0}\ \widehat{a_1}\ \dots\ \widehat{a_p})'$$

throughout this chapter without further notice.

Lemma 5.4

Suppose that X_t is generated by the model (5.1) satisfying assumptions (A1)-(A2) and (B1)-(B4). Then for all $e \in \mathbb{R}$

$$\sqrt{\frac{h}{T}} \sum_{i=1}^{T} K_h(e - \widehat{\varepsilon_{i-1}})\widehat{\varepsilon_i}^2 = \sqrt{\frac{h}{T}} \sum_{i=1}^{T} K_h(e - \varepsilon_{i-1})\varepsilon_i^2 + o_p(1).$$

Proof. Here it suffices to show

$$\sqrt{\frac{h}{T}} \sum_{i=1}^{T} \left(K_h(e - \widehat{\varepsilon_{i-1}}) - K_h(e - \varepsilon_{i-1})\right)\varepsilon_i^2 = o_p(1) \tag{5.3}$$

and

$$\sqrt{\frac{h}{T}} \sum_{i=1}^{T} K_h(e - \widehat{\varepsilon_{i-1}})\left(\widehat{\varepsilon_i}^2 - \varepsilon_i^2\right) = o_p(1). \tag{5.4}$$

Firstly, a Taylor expansion for the left hand side of (5.3) yields

$$
\begin{aligned}
&\sqrt{\frac{h}{T}}\sum_{i=1}^{T}\left(K_h(e-\widehat{\varepsilon_{i-1}})-K_h(e-\varepsilon_{i-1})\right)\varepsilon_i^2\\
=&\frac{1}{\sqrt{Th}}\sum_{i=1}^{T}\left(K\left(\frac{e-\widehat{\varepsilon_{i-1}}}{h}\right)-K\left(\frac{e-\varepsilon_{i-1}}{h}\right)\right)\varepsilon_i^2\\
=&\frac{1}{\sqrt{Th}}\sum_{i=1}^{T}\left(K'\left(\frac{e-\varepsilon_{i-1}}{h}\right)\frac{\varepsilon_{i-1}-\widehat{\varepsilon_{i-1}}}{h}\right.\\
&\left.+\frac{1}{2}K''\left(\frac{e-\varepsilon_{i-1}}{h}\right)\left(\frac{\varepsilon_{i-1}-\widehat{\varepsilon_{i-1}}}{h}\right)^2+\frac{1}{6}K'''(\widetilde{e})\left(\frac{\varepsilon_{i-1}-\widehat{\varepsilon_{i-1}}}{h}\right)^3\right)\varepsilon_i^2\\
=&\frac{1}{\sqrt{Th^3}}\sum_{i=1}^{T}\varepsilon_i^2K'\left(\frac{e-\varepsilon_{i-1}}{h}\right)\mathbf{x_{i-1}}'(\widehat{\mathbf{a}}-\mathbf{a})\\
&+\frac{1}{2\sqrt{Th^5}}\sum_{i=1}^{T}\varepsilon_i^2K''\left(\frac{e-\varepsilon_{i-1}}{h}\right)\left(\mathbf{x_{i-1}}'(\widehat{\mathbf{a}}-\mathbf{a})\right)^2\\
&+\frac{1}{6\sqrt{Th^7}}\sum_{i=1}^{T}\varepsilon_i^2K'''(\widetilde{e})\left(\mathbf{x_{i-1}}'(\widehat{\mathbf{a}}-\mathbf{a})\right)^3\\
\le&\left(\frac{1}{T}\sum_{i=1}^{T}\frac{\varepsilon_i^2}{\sqrt{h^3}}K'\left(\frac{e-\varepsilon_{i-1}}{h}\right)\mathbf{x_{i-1}}'\right)\underbrace{\sqrt{T}(\widehat{\mathbf{a}}-\mathbf{a})}_{=O_p(1)}\\
&+\frac{1}{2}\left(\frac{1}{T}\sum_{i=1}^{T}\frac{\varepsilon_i^2}{\sqrt{Th^5}}K''\left(\frac{e-\varepsilon_{i-1}}{h}\right)\sum_{k=0}^{p}x_{i-1,k}^2\right)\underbrace{T\sum_{k=0}^{p}(\widehat{a_k}-a_k)^2}_{=O_p(1)}\\
&+\frac{c_p}{6\sqrt{T^2h^7}}\underbrace{\max_{u\in\mathbb{R}}\left|K'''(u)\right|}_{=O(1)}\left(\frac{1}{T}\sum_{i=1}^{T}\underbrace{\varepsilon_i^2\sum_{k=0}^{p}|x_{i-1,k}|^3}_{=O_p(1)}\right)\underbrace{T^{3/2}\sum_{k=0}^{p}|\widehat{a_k}-a_k|^3}_{=O_p(1)},
\end{aligned}
$$

where $\widetilde{e}$ denotes a suitable value between $\frac{e-\widehat{\varepsilon_{i-1}}}{h}$ and $\frac{e-\varepsilon_{i-1}}{h}$, and c_p a constant such that $0<c_p<\infty$.

Now let

$$\mathbf{z_{i-1}}':=\frac{\varepsilon_i^2}{\sqrt{h^3}}K'\left(\frac{e-\varepsilon_{i-1}}{h}\right)\mathbf{x_{i-1}}',$$

then on the one hand

$$
\begin{aligned}
&\frac{1}{T}\mathscr{E}\left(\mathbf{z_{i-1}}'\mathbf{z_{i-1}}\right)\\
=&\frac{1}{T}\mathscr{E}\left\{\frac{\mathbf{x_{i-1}}'\mathbf{x_{i-1}}\varepsilon_i^4}{h^3}\left(K'\right)^2\left(\frac{e-\varepsilon_{i-1}}{h}\right)\right\}\\
\leq&\frac{1}{Th^3}\underbrace{\max_{u\in\mathbb{R}}\left(K'\right)^2(u)}_{=O(1)}\underbrace{\mathscr{E}\left(\mathbf{x_{i-1}}'\mathbf{x_{i-1}}\varepsilon_i^4\right)}_{=O(1)}\\
=&o(1),
\end{aligned}
$$

thus Remark B.1 yields

$$
\frac{1}{T}\sum_{i=1}^{T}\mathbf{z_{i-1}}' = \frac{1}{T}\sum_{i=1}^{T}\mathscr{E}\left(\mathbf{z_{i-1}}'\middle|\mathscr{F}_{i-2}\right)+o_p(1).
$$

On the other hand, because $\mathbf{x_{i-1}}$ is measurable to $\mathscr{F}_{i-2}$, we obtain

$$
\begin{aligned}
&\frac{1}{T}\sum_{i=1}^{T}\mathscr{E}\left(\mathbf{z_{i-1}}'\middle|\mathscr{F}_{i-2}\right)\\
=&\frac{1}{T}\sum_{i=1}^{T}\mathscr{E}\left(\frac{\mathbf{x_{i-1}}'\varepsilon_i^2}{\sqrt{h^3}}K'\left(\frac{e-\varepsilon_{i-1}}{h}\right)\middle|\mathscr{F}_{i-2}\right)\\
=&\frac{1}{T}\sum_{i=1}^{T}\frac{\mathbf{x_{i-1}}'}{\sqrt{h^3}}\mathscr{E}\left(\sigma^2(\varepsilon_{i-1})K'\left(\frac{e-\varepsilon_{i-1}}{h}\right)\middle|\mathscr{F}_{i-2}\right)\\
=&\frac{1}{T}\sum_{i=1}^{T}\frac{\mathbf{x_{i-1}}'}{\sqrt{h^3}}\int\sigma^2(u)K'\left(\frac{e-u}{h}\right)p(u)du\\
=&\frac{1}{T}\sum_{i=1}^{T}\frac{\mathbf{x_{i-1}}'}{\sqrt{h^3}}\int\sigma^2(e-hv)K'(v)p(e-hv)(-h)dv\\
=&-\frac{1}{T}\sum_{i=1}^{T}\frac{\mathbf{x_{i-1}}'}{\sqrt{h}}\int\sigma^2(e-hv)K'(v)p(e-hv)dv\\
=&-\frac{1}{T}\sum_{i=1}^{T}\frac{\mathbf{x_{i-1}}'}{\sqrt{h}}\int\left(\sigma^2(e)-\left(\sigma^2\right)'(e)hv+\frac{1}{2}\left(\sigma^2\right)''(\widetilde{e})h^2v^2\right)\\
&K'(v)\left(p(e)-p'(e)hv+\frac{1}{2}p''(\widetilde{e})h^2v^2\right)dv\\
=&-\frac{1}{\sqrt{h}}\sigma^2(e)p(e)\underbrace{\int K'(v)dv}_{=0}\left(\frac{1}{T}\sum_{i=1}^{T}\mathbf{x_{i-1}}'\right)\\
&+\sqrt{h}\left(\sigma^2(e)p'(e)+\left(\sigma^2\right)'(e)p(e)\right)\underbrace{\int vK'(v)dv}_{=-1}\left(\frac{1}{T}\sum_{i=1}^{T}\mathbf{x_{i-1}}'\right)\\
&-\sqrt{h^3}\left(\frac{1}{2}\sigma^2(e)p''(\widetilde{e})+\left(\sigma^2\right)'(e)p'(e)+\frac{1}{2}\left(\sigma^2\right)''(\widetilde{e})p(e)\right)\\
&\underbrace{\int v^2K'(v)dv}_{=O(1)}\left(\frac{1}{T}\sum_{i=1}^{T}\mathbf{x_{i-1}}'\right)\\
&+\frac{\sqrt{h^5}}{2}\left(\left(\sigma^2\right)'(e)p''(\widetilde{e})+\left(\sigma^2\right)''(\widetilde{e})p'(e)\right)\underbrace{\int v^3K'(v)dv}_{=O(1)}\left(\frac{1}{T}\sum_{i=1}^{T}\mathbf{x_{i-1}}'\right)\\
&-\frac{\sqrt{h^7}}{4}\left(\sigma^2\right)''(\widetilde{e})p''(\widetilde{e})\underbrace{\int v^4K'(v)dv}_{=O(1)}\left(\frac{1}{T}\sum_{i=1}^{T}\mathbf{x_{i-1}}'\right)\\
=&o_p(1),
\end{aligned}
$$

and, therefore, we obtain

$$\frac{1}{T}\sum_{i=1}^{T}\mathbf{z_{i-1}}' = o_p(1).$$

Analogously, we can prove

$$\frac{1}{T}\sum_{i=1}^{T}\frac{\varepsilon_i^2}{\sqrt{Th^5}}K''\left(\frac{e-\varepsilon_{i-1}}{h}\right)\sum_{k=0}^{p}x_{i-1,k}^2 = o_p(1),$$

which shows (5.3).

Secondly, the left hand side of (5.4) equals

$$\begin{aligned}
&\sqrt{\frac{h}{T}}\sum_{i=1}^{T}K_h(e-\widehat{\varepsilon_{i-1}})\left(\widehat{\varepsilon_i^2}-\varepsilon_i^2\right)\\
=&\frac{1}{\sqrt{Th}}\sum_{i=1}^{T}K\left(\frac{e-\widehat{\varepsilon_{i-1}}}{h}\right)\left(\widehat{\varepsilon_i^2}-\varepsilon_i^2\right)\\
\leq&\max_{u\in\mathbb{R}}\left|K(u)\right|\frac{1}{\sqrt{Th}}\sum_{i=1}^{T}\left|\widehat{\varepsilon_i^2}-\varepsilon_i^2\right|\\
\leq&\max_{u\in\mathbb{R}}\left|K(u)\right|\left(\frac{1}{\sqrt{Th}}\sum_{i=1}^{T}\left(\widehat{\varepsilon_i}-\varepsilon_i\right)^2+\frac{2}{\sqrt{Th}}\sum_{i=1}^{T}\left|\varepsilon_i\left(\widehat{\varepsilon_i}-\varepsilon_i\right)\right|\right)\\
=&\max_{u\in\mathbb{R}}\left|K(u)\right|\left(\frac{1}{\sqrt{Th}}\sum_{i=1}^{T}\left(\mathbf{x_i}'\left(\mathbf{a}-\widehat{\mathbf{a}}\right)\right)^2+\frac{2}{\sqrt{Th}}\sum_{i=1}^{T}\left|\varepsilon_i\mathbf{x_i}'\left(\mathbf{a}-\widehat{\mathbf{a}}\right)\right|\right)\\
\leq&\max_{u\in\mathbb{R}}\left|K(u)\right|\frac{1}{\sqrt{Th}}\underbrace{\left(T\sum_{k=0}^{p}(a_k-\widehat{a_k})^2\right)}_{=O_p(1)}\left(\frac{1}{T}\sum_{i=1}^{T}\underbrace{\sum_{k=0}^{p}x_{i,k}^2}_{=O_p(1)}\right)\\
&+\max_{u\in\mathbb{R}}\left|K(u)\right|\underbrace{\left(\frac{2}{\sqrt{T^2h}}\sum_{i=1}^{T}\varepsilon_i\mathbf{x_i}'\right)}_{=o_p(1)}\underbrace{\sqrt{T}\left(\mathbf{a}-\widehat{\mathbf{a}}\right)}_{=O_p(1)}\\
=&o_p(1).
\end{aligned}$$

Here we obtain

$$\mathbf{y_t}' := \frac{2}{\sqrt{T^2h}}\sum_{i=1}^{T}\varepsilon_i\mathbf{x_i}' = o_p(1)$$

because

$$\mathscr{E}\left(\mathbf{y_t}'\right) = \frac{2}{\sqrt{T^2h}} \sum_{i=1}^{T} \underbrace{\mathscr{E}(\eta_i)}_{=0} \mathscr{E}\left(\sigma(\varepsilon_{i-1})\mathbf{x_i}'\right) = 0,$$

$$\begin{aligned}\mathscr{E}\left(\mathbf{y_t}'\mathbf{y_t}\right) &= \frac{4}{T^2h} \sum_{i,j=1}^{T} \mathscr{E}\left(\varepsilon_i\varepsilon_j\mathbf{x_i}'\mathbf{x_j}\right) \\ &= \frac{4}{Th}\left(\frac{1}{T}\sum_{i=1}^{T}\mathscr{E}\left(\varepsilon_i^2\mathbf{x_i}'\mathbf{x_i}\right)\right) \\ &= o(1),\end{aligned}$$

and, therefore, from Chebyshev's inequality we obtain for every $\delta > 0$

$$P\left\{\left|\mathbf{y_t}' - \mathscr{E}\left(\mathbf{y_t}'\right)\right| > \delta\right\} \leq \frac{1}{\delta^2}\mathscr{E}\left(\mathbf{y_t}'\mathbf{y_t}\right) = o(1),$$

that is, according to the definition $\mathbf{y_t}' \xrightarrow{p} \mathscr{E}\left(\mathbf{y_t}'\right) = 0$. □

Lemma 5.5

Suppose that X_t is generated by the model (5.1) satisfying assumptions (A1)-(A2) and (B1)-(B4). Then for all $e \in \mathbb{R}$

$$\frac{1}{T}\sum_{j=1}^{T} K_h(e - \widehat{\varepsilon_{j-1}}) = \frac{1}{T}\sum_{j=1}^{T} K_h(e - \varepsilon_{j-1}) + o_p\left(\frac{1}{\sqrt{Th}}\right).$$

Proof. Analogously to Lemma 5.4, we prove

$$\sqrt{\frac{h}{T}}\sum_{i=1}^{T}\left(K_h(e - \widehat{\varepsilon_{i-1}}) - K_h(e - \varepsilon_{i-1})\right) = o_p(1).$$

A Taylor expansion for the left hand side yields

$$\begin{aligned}
&\sqrt{\frac{h}{T}}\sum_{i=1}^{T}\Big(K_h(e-\widehat{\varepsilon_{i-1}})-K_h(e-\varepsilon_{i-1})\Big)\\
=&\frac{1}{\sqrt{Th}}\sum_{i=1}^{T}\left(K\left(\frac{e-\widehat{\varepsilon_{i-1}}}{h}\right)-K\left(\frac{e-\varepsilon_{i-1}}{h}\right)\right)\\
=&\frac{1}{\sqrt{Th}}\sum_{i=1}^{T}\left(K'\left(\frac{e-\varepsilon_{i-1}}{h}\right)\frac{\varepsilon_{i-1}-\widehat{\varepsilon_{i-1}}}{h}\right.\\
&\left.+\frac{1}{2}K''\left(\frac{e-\varepsilon_{i-1}}{h}\right)\left(\frac{\varepsilon_{i-1}-\widehat{\varepsilon_{i-1}}}{h}\right)^2+\frac{1}{6}K'''(\widetilde{e})\left(\frac{\varepsilon_{i-1}-\widehat{\varepsilon_{i-1}}}{h}\right)^3\right)\\
=&\frac{1}{\sqrt{Th^3}}\sum_{i=1}^{T}K'\left(\frac{e-\varepsilon_{i-1}}{h}\right)\mathbf{x_{i-1}}'(\widehat{\mathbf{a}}-\mathbf{a})\\
&+\frac{1}{2\sqrt{Th^5}}\sum_{i=1}^{T}K''\left(\frac{e-\varepsilon_{i-1}}{h}\right)\Big(\mathbf{x_{i-1}}'(\widehat{\mathbf{a}}-\mathbf{a})\Big)^2\\
&+\frac{1}{6\sqrt{Th^7}}\sum_{i=1}^{T}K'''(\widetilde{e})\Big(\mathbf{x_{i-1}}'(\widehat{\mathbf{a}}-\mathbf{a})\Big)^3\\
\le&\underbrace{\left(\frac{1}{T}\sum_{i=1}^{T}\frac{1}{\sqrt{h^3}}K'\left(\frac{e-\varepsilon_{i-1}}{h}\right)\mathbf{x_{i-1}}'\right)}_{=o_p(1)}\underbrace{\sqrt{T}(\widehat{\mathbf{a}}-\mathbf{a})}_{=O_p(1)}\\
&+\frac{1}{2}\underbrace{\left(\frac{1}{T}\sum_{i=1}^{T}\frac{1}{\sqrt{Th^5}}K''\left(\frac{e-\varepsilon_{i-1}}{h}\right)\sum_{k=0}^{p}x_{i-1,k}^2\right)}_{=o_p(1)}\underbrace{T\sum_{k=0}^{p}(\widehat{a_k}-a_k)^2}_{=O_p(1)}\\
&+\frac{c_p}{6\sqrt{T^2h^7}}\underbrace{\max_{u\in\mathbb{R}}\left|K'''(u)\right|}_{=O(1)}\left(\frac{1}{T}\sum_{i=1}^{T}\underbrace{\sum_{k=0}^{p}|x_{i-1,k}|^3}_{=O_p(1)}\right)\underbrace{T^{3/2}\sum_{k=0}^{p}|\widehat{a_k}-a_k|^3}_{=O_p(1)}\\
=&o_p(1),
\end{aligned}$$

where $\widetilde{e}$ denotes a suitable value between $\frac{e-\widehat{\varepsilon_{i-1}}}{h}$ and $\frac{e-\varepsilon_{i-1}}{h}$, and c_p a constant such that $0<c_p<\infty$. □

Theorem 5.2

Suppose that X_t is generated by the model (5.1) satisfying assumptions (A1)-

(A2) and (B1)-(B4). Then for all $e \in \mathbb{R}$

$$\sqrt{Th}\left(\widetilde{\sigma_h^2}(e) - \sigma^2(e)\right) \xrightarrow{d} \mathscr{N}\left(\frac{b(e)}{p(e)}, \frac{\tau^2(e)}{p^2(e)}\right).$$

Proof. The left hand side equals

$$\sqrt{Th}\left(\widetilde{\sigma_h^2}(e) - \sigma^2(e)\right) = \frac{\sqrt{\frac{h}{T}}\sum_{i=1}^T K_h(e - \widehat{\varepsilon_{i-1}})\left(\widehat{\varepsilon_i}^2 - \sigma^2(e)\right)}{\frac{1}{T}\sum_{j=1}^T K_h(e - \widehat{\varepsilon_{j-1}})}.$$

From Lemma 5.1-5.3 we already know

$$\sqrt{\frac{h}{T}}\sum_{i=1}^T K_h(e - \varepsilon_{i-1})\left(\varepsilon_i^2 - \sigma^2(e)\right) \xrightarrow{d} \mathscr{N}\left(b(e), \tau^2(e)\right) \quad \textit{and}$$

$$\frac{1}{T}\sum_{j=1}^T K_h(e - \varepsilon_{j-1}) \xrightarrow{p} p(e).$$

From Lemma 5.5 we already have

$$\frac{1}{T}\sum_{j=1}^T K_h(e - \widehat{\varepsilon_{j-1}}) = \frac{1}{T}\sum_{j=1}^T K_h(e - \varepsilon_{j-1}) + o_p(1).$$

Together with the Slutsky theorem it suffices to show

$$\begin{aligned}&\sqrt{\frac{h}{T}}\sum_{i=1}^T K_h(e - \widehat{\varepsilon_{i-1}})\left(\widehat{\varepsilon_i}^2 - \sigma^2(e)\right)\\ =&\sqrt{\frac{h}{T}}\sum_{i=1}^T K_h(e - \varepsilon_{i-1})\left(\varepsilon_i^2 - \sigma^2(e)\right) + o_p(1).\end{aligned}$$

From Lemma 5.4 and 5.5 we obtain

$$\begin{aligned}
&\sqrt{\frac{h}{T}}\sum_{i=1}^{T} K_h(e-\widehat{\varepsilon_{i-1}})\left(\widehat{\varepsilon_i}^2-\sigma^2(e)\right)\\
=&\sqrt{\frac{h}{T}}\sum_{i=1}^{T} K_h(e-\widehat{\varepsilon_{i-1}})\widehat{\varepsilon_i}^2-\sqrt{Th}\sigma^2(e)\frac{1}{T}\sum_{i=1}^{T} K_h(e-\widehat{\varepsilon_{i-1}})\\
=&\sqrt{\frac{h}{T}}\sum_{i=1}^{T} K_h(e-\varepsilon_{i-1})\varepsilon_i^2+o_p(1)\\
&-\sqrt{Th}\sigma^2(e)\left(\frac{1}{T}\sum_{i=1}^{T} K_h(e-\varepsilon_{i-1})+o_p\left(\frac{1}{\sqrt{Th}}\right)\right)\\
=&\sqrt{\frac{h}{T}}\sum_{i=1}^{T} K_h(e-\varepsilon_{i-1})\left(\varepsilon_i^2-\sigma^2(e)\right)+o_p(1)-o_p(1).
\end{aligned}$$

□

5.2 Residual Bootstrap

Franke, Kreiß and Mammen (2002) proved asymptotic validity of the residual, the wild and the autoregression bootstrap technique applied to the NW estimators for the NARCH(1) model. In this and the following section we introduce possible ways to bootstrap the semiparametric AR(p)-ARCH(1) model based on the results of Franke, Kreiß and Mammen (2002).

In this section a residual bootstrap technique is proposed and its weak consistency proved. A residual bootstrap method can be applied to the model (5.1) as follows:

(step 1) Obtain the OLS estimator $\widehat{\mathbf{a}}$ and calculate the residuals

$$\widehat{\varepsilon_t}=X_t-\widehat{a_0}-\sum_{i=1}^{p}\widehat{a_i}X_{t-i},\qquad t=0,\dots,T.$$

(step 2) Obtain the NW estimator with a bandwidth g

$$\widetilde{\sigma_g^2}(e)=\frac{T^{-1}\sum_{i=1}^{T}K_g(e-\widehat{\varepsilon_{i-1}})\widehat{\varepsilon_i}^2}{T^{-1}\sum_{j=1}^{T}K_g(e-\widehat{\varepsilon_{j-1}})}.$$

(step 3) Compute the estimated bias

$$\widehat{m_t} = \left(\frac{\widehat{\varepsilon_t}^2}{\widetilde{\sigma_g^2}(\widehat{\varepsilon_{t-1}})} - 1 \right) (\kappa - 1)^{-1/2}$$

for $t = 1, ..., T$ and the standardised estimated bias

$$\widetilde{m_t} = \frac{\widehat{m_t} - \widehat{\mu}}{\widehat{\sigma}}$$

for $t = 1, ..., T$, where

$$\widehat{\mu} = \frac{1}{T} \sum_{t=1}^{T} \widehat{m_t} \qquad \textit{and} \qquad \widehat{\sigma}^2 = \frac{1}{T} \sum_{t=1}^{T} (\widehat{m_t} - \widehat{\mu})^2.$$

(step 4) Obtain the empirical distribution function $\mathscr{F}_T(x)$ based on $\widetilde{m_t}$ defined by

$$\mathscr{F}_T(x) := \frac{1}{T} \sum_{t=1}^{T} \mathbf{1}(\widetilde{m_t} \leq x).$$

(step 5) Generate the bootstrap process ε_t^{*2} by computing

$$\varepsilon_t^{*2} = \widetilde{\sigma_g^2}(\widehat{\varepsilon_{t-1}}) + \sqrt{\kappa - 1}\widetilde{\sigma_g^2}(\widehat{\varepsilon_{t-1}}) m_t^*,$$
$$m_t^* \overset{iid}{\sim} \mathscr{F}_T(x), \qquad t = 1, ..., T.$$

(step 6) Build the bootstrap model

$$\varepsilon_t^{*2} = \sigma^{*2}(\widehat{\varepsilon_{t-1}}) + \sqrt{\kappa - 1}\sigma^{*2}(\widehat{\varepsilon_{t-1}}) m_t^*$$

and calculate the NW estimator $\widetilde{\sigma_h^{*2}}(e)$ with another bandwidth h

$$\widetilde{\sigma_h^{*2}}(e) = \frac{T^{-1} \sum_{i=1}^{T} K_h(e - \widehat{\varepsilon_{i-1}}) \varepsilon_i^{*2}}{T^{-1} \sum_{j=1}^{T} K_h(e - \widehat{\varepsilon_{j-1}})}.$$

Analogously to the previous section, we observe the decomposition

$$\begin{aligned}
&\sqrt{Th}\left(\widetilde{\sigma_h^{*2}}(e) - \widetilde{\sigma_g^2}(e)\right) \\
&= \frac{\sqrt{\frac{h}{T}} \sum_{i=1}^{T} K_h(e - \widehat{\varepsilon_{i-1}}) \sqrt{\kappa - 1}\widetilde{\sigma_g^2}(\widehat{\varepsilon_{i-1}}) m_i^*}{\frac{1}{T} \sum_{j=1}^{T} K_h(e - \widehat{\varepsilon_{j-1}})} \\
&\quad + \frac{\sqrt{\frac{h}{T}} \sum_{i=1}^{T} K_h(e - \widehat{\varepsilon_{i-1}}) \left(\widetilde{\sigma_g^2}(\widehat{\varepsilon_{i-1}}) - \widetilde{\sigma_g^2}(e)\right)}{\frac{1}{T} \sum_{j=1}^{T} K_h(e - \widehat{\varepsilon_{j-1}})}
\end{aligned}$$

and analyse the asymptotic properties of each part separately.

For the proof of weak consistency of the residual bootstrap we need a stronger condition than just to prove asymptotic normality of the NW estimators.

(B5) $\mathscr{E}|\eta_t|^6 < \infty$.

Lemma 5.6

Suppose that X_t is generated by the model (5.1) satisfying assumptions (A1)-(A2) and (B1)-(B4). Then for all $e \in \mathbb{R}$

$$\left(\widetilde{\sigma_g^2}\right)'(e) \xrightarrow{p} \left(\sigma^2\right)'(e) \quad and \quad \left(\widetilde{\sigma_g^2}\right)''(e) \xrightarrow{p} \left(\sigma^2\right)''(e).$$

Proof. We observe the decomposition

$$\begin{aligned}\left(\widetilde{\sigma_g^2}\right)'(e) =& \frac{\frac{1}{Tg^2}\sum_{i=1}^T K'\left(\frac{e-\widehat{\varepsilon_{i-1}}}{g}\right)\widehat{\varepsilon_i}^2}{\frac{1}{Tg}\sum_{j=1}^T K\left(\frac{e-\widehat{\varepsilon_{j-1}}}{g}\right)} \\ &- \frac{\frac{1}{Tg}\sum_{i=1}^T K\left(\frac{e-\widehat{\varepsilon_{i-1}}}{g}\right)\widehat{\varepsilon_i}^2\left(\frac{1}{Tg^2}\sum_{i=1}^T K'\left(\frac{e-\widehat{\varepsilon_{i-1}}}{g}\right)\right)}{\left(\frac{1}{Tg}\sum_{j=1}^T K\left(\frac{e-\widehat{\varepsilon_{j-1}}}{g}\right)\right)^2}\end{aligned}$$

and

$$\begin{aligned}\left(\widetilde{\sigma_g^2}\right)''(e) =& \frac{\frac{1}{Tg^3}\sum_{i=1}^T K''\left(\frac{e-\widehat{\varepsilon_{i-1}}}{g}\right)\widehat{\varepsilon_i}^2}{\frac{1}{Tg}\sum_{j=1}^T K\left(\frac{e-\widehat{\varepsilon_{j-1}}}{g}\right)} \\ &-2\frac{\frac{1}{Tg^2}\sum_{i=1}^T K'\left(\frac{e-\widehat{\varepsilon_{i-1}}}{g}\right)\widehat{\varepsilon_i}^2}{\left(\frac{1}{Tg}\sum_{j=1}^T K\left(\frac{e-\widehat{\varepsilon_{j-1}}}{g}\right)\right)^2} \\ &- \frac{\frac{1}{Tg}\sum_{i=1}^T K\left(\frac{e-\widehat{\varepsilon_{i-1}}}{g}\right)\widehat{\varepsilon_i}^2\left(\frac{1}{Tg^3}\sum_{i=1}^T K''\left(\frac{e-\widehat{\varepsilon_{i-1}}}{g}\right)\right)}{\left(\frac{1}{Tg}\sum_{j=1}^T K\left(\frac{e-\widehat{\varepsilon_{j-1}}}{g}\right)\right)^2} \\ &+2\frac{\frac{1}{Tg}\sum_{i=1}^T K\left(\frac{e-\widehat{\varepsilon_{i-1}}}{g}\right)\widehat{\varepsilon_i}^2\left(\frac{1}{Tg^2}\sum_{i=1}^T K'\left(\frac{e-\widehat{\varepsilon_{i-1}}}{g}\right)\right)}{\left(\frac{1}{Tg}\sum_{j=1}^T K\left(\frac{e-\widehat{\varepsilon_{j-1}}}{g}\right)\right)^3}.\end{aligned}$$

Together with the Slutsky theorem it suffices to show

$$(i)\quad \frac{1}{Tg^2}\sum_{i=1}^{T}\left(K'\left(\frac{e-\varepsilon_{i-1}}{g}\right)-K'\left(\frac{e-\widehat{\varepsilon_{i-1}}}{g}\right)\right)=o_p(1),$$

$$(ii)\quad \frac{1}{Tg^2}\sum_{i=1}^{T}\left(K'\left(\frac{e-\varepsilon_{i-1}}{g}\right)\varepsilon_i^2-K'\left(\frac{e-\widehat{\varepsilon_{i-1}}}{g}\right)\widehat{\varepsilon_i}^2\right)=o_p(1),$$

$$(iii)\quad \frac{1}{Tg^3}\sum_{i=1}^{T}\left(K''\left(\frac{e-\varepsilon_{i-1}}{g}\right)-K''\left(\frac{e-\widehat{\varepsilon_{i-1}}}{g}\right)\right)=o_p(1),$$

$$(iv)\quad \frac{1}{Tg^3}\sum_{i=1}^{T}\left(K''\left(\frac{e-\varepsilon_{i-1}}{g}\right)\varepsilon_i^2-K''\left(\frac{e-\widehat{\varepsilon_{i-1}}}{g}\right)\widehat{\varepsilon_i}^2\right)=o_p(1).$$

A Taylor expansion for the left hand side of (i) yields

$$\begin{aligned}
&\frac{1}{Tg^2}\sum_{i=1}^{T}\left(K'\left(\frac{e-\widehat{\varepsilon_{i-1}}}{g}\right)-K'\left(\frac{e-\varepsilon_{i-1}}{g}\right)\right)\\
=&\frac{1}{Tg^2}\sum_{i=1}^{T}\left(K''\left(\frac{e-\varepsilon_{i-1}}{g}\right)\frac{\varepsilon_{i-1}-\widehat{\varepsilon_{i-1}}}{g}+\frac{1}{2}K'''(\widetilde{e})\left(\frac{\varepsilon_{i-1}-\widehat{\varepsilon_{i-1}}}{g}\right)^2\right)\\
=&\frac{1}{Tg^3}\sum_{i=1}^{T}K''\left(\frac{e-\varepsilon_{i-1}}{g}\right)\mathbf{x_{i-1}}'(\widehat{\mathbf{a}}-\mathbf{a})+\frac{1}{2Tg^4}\sum_{i=1}^{T}K'''(\widetilde{e})\left(\mathbf{x_{i-1}}'(\widehat{\mathbf{a}}-\mathbf{a})\right)^2\\
\leq&\underbrace{\left(\frac{1}{T}\sum_{i=1}^{T}\frac{1}{\sqrt{Tg^6}}K''\left(\frac{e-\varepsilon_{i-1}}{g}\right)\mathbf{x_{i-1}}'\right)}_{=o_p(1)}\underbrace{\sqrt{T}(\widehat{\mathbf{a}}-\mathbf{a})}_{=O_p(1)}\\
&+\frac{1}{2}\underbrace{\left(\frac{1}{T}\sum_{i=1}^{T}\frac{1}{Tg^4}K'''(\widetilde{e})\sum_{k=0}^{p}x_{i-1,k}^2\right)}_{=o_p(1)}\underbrace{T\sum_{k=0}^{p}(\widehat{a_k}-a_k)^2}_{=O_p(1)}\\
=&o_p(1),
\end{aligned}$$

where $\widetilde{e}$ denotes a suitable value between $\frac{e-\widehat{\varepsilon_{i-1}}}{g}$ and $\frac{e-\varepsilon_{i-1}}{g}$.

To prove (ii) we need to show

$$\frac{1}{Tg^2}\sum_{i=1}^{T}\left(K'\left(\frac{e-\widehat{\varepsilon_{i-1}}}{g}\right)-K'\left(\frac{e-\varepsilon_{i-1}}{g}\right)\right)\varepsilon_i^2=o_p(1) \tag{5.5}$$

and

$$\frac{1}{Tg^2}\sum_{i=1}^{T} K'\left(\frac{e-\widehat{\varepsilon_{i-1}}}{g}\right)\left(\widehat{\varepsilon_i}^2-\varepsilon_i^2\right)=o_p(1). \tag{5.6}$$

Firstly, a Taylor expansion for the left hand side of (5.5) yields

$$\begin{aligned}
&\frac{1}{Tg^2}\sum_{i=1}^{T}\left(K'\left(\frac{e-\widehat{\varepsilon_{i-1}}}{g}\right)-K'\left(\frac{e-\varepsilon_{i-1}}{g}\right)\right)\varepsilon_i^2\\
=&\frac{1}{Tg^2}\sum_{i=1}^{T}\left(K''\left(\frac{e-\varepsilon_{i-1}}{g}\right)\frac{\varepsilon_{i-1}-\widehat{\varepsilon_{i-1}}}{g}+\frac{1}{2}K'''(\widetilde{e})\left(\frac{\varepsilon_{i-1}-\widehat{\varepsilon_{i-1}}}{g}\right)^2\right)\varepsilon_i^2\\
=&\frac{1}{Tg^3}\sum_{i=1}^{T}\varepsilon_i^2K''\left(\frac{e-\varepsilon_{i-1}}{g}\right)\mathbf{x_{i-1}}'\left(\widehat{\mathbf{a}}-\mathbf{a}\right)\\
&+\frac{1}{2Tg^4}\sum_{i=1}^{T}\varepsilon_i^2K'''(\widetilde{e})\left(\mathbf{x_{i-1}}'\left(\widehat{\mathbf{a}}-\mathbf{a}\right)\right)^2\\
\le&\underbrace{\left(\frac{1}{T}\sum_{i=1}^{T}\frac{\varepsilon_i^2}{\sqrt{Tg^6}}K''\left(\frac{e-\varepsilon_{i-1}}{g}\right)\mathbf{x_{i-1}}'\right)}_{=o_p(1)}\underbrace{\sqrt{T}\left(\widehat{\mathbf{a}}-\mathbf{a}\right)}_{=O_p(1)}\\
&+\frac{1}{2}\underbrace{\left(\frac{1}{T}\sum_{i=1}^{T}\frac{\varepsilon_i^2}{Tg^4}K'''(\widetilde{e})\sum_{k=0}^{p}x_{i-1,k}^2\right)}_{=o_p(1)}\underbrace{T\sum_{k=0}^{p}\left(\widehat{a_k}-a_k\right)^2}_{=O_p(1)}\\
=&o_p(1),
\end{aligned}$$

where $\widetilde{e}$ denotes a suitable value between $\frac{e-\widehat{\varepsilon_{i-1}}}{g}$ and $\frac{e-\varepsilon_{i-1}}{g}$. Secondly, the left hand side of (5.6) equals

$$
\begin{aligned}
&\frac{1}{Tg^2}\sum_{i=1}^{T} K'\left(\frac{e-\widehat{\varepsilon_{i-1}}}{g}\right)\left(\widehat{\varepsilon_i^2}-\varepsilon_i^2\right)\\
\leq &\max_{u\in\mathbb{R}}\left|K'(u)\right|\frac{1}{Tg^2}\sum_{i=1}^{T}\left|\widehat{\varepsilon_i^2}-\varepsilon_i^2\right|\\
\leq &\max_{u\in\mathbb{R}}\left|K'(u)\right|\left(\frac{1}{Tg^2}\sum_{i=1}^{T}\left(\widehat{\varepsilon}_i-\varepsilon_i\right)^2+\frac{2}{Tg^2}\sum_{i=1}^{T}\left|\varepsilon_i\left(\widehat{\varepsilon}_i-\varepsilon_i\right)\right|\right)\\
= &\max_{u\in\mathbb{R}}\left|K'(u)\right|\left(\frac{1}{Tg^2}\sum_{i=1}^{T}\left(\mathbf{x_i}'\left(\mathbf{a}-\widehat{\mathbf{a}}\right)\right)^2+\frac{2}{Tg^2}\sum_{i=1}^{T}\left|\varepsilon_i\mathbf{x_i}'\left(\mathbf{a}-\widehat{\mathbf{a}}\right)\right|\right)\\
\leq &\max_{u\in\mathbb{R}}\left|K'(u)\right|\frac{1}{Tg^2}\underbrace{\left(T\sum_{k=0}^{p}\left(a_k-\widehat{a_k}\right)^2\right)}_{=O_p(1)}\left(\frac{1}{T}\sum_{i=1}^{T}\underbrace{\sum_{k=0}^{p}x_{i,k}^2}_{=O_p(1)}\right)\\
&+\max_{u\in\mathbb{R}}\left|K'(u)\right|\underbrace{\left(\frac{2}{\sqrt{T^3g^4}}\sum_{i=1}^{T}\left|\varepsilon_i\mathbf{x_i}'\right|\right)}_{=o_p(1)}\underbrace{\sqrt{T}\left|\mathbf{a}-\widehat{\mathbf{a}}\right|}_{=O_p(1)}\\
= &o_p(1).
\end{aligned}
$$

A Taylor expansion for the left hand side of (iii) yields

$$
\begin{aligned}
&\frac{1}{Tg^3}\sum_{i=1}^{T}\left(K''\left(\frac{e-\widehat{\varepsilon_{i-1}}}{g}\right)-K''\left(\frac{e-\varepsilon_{i-1}}{g}\right)\right)\\
= &\frac{1}{Tg^3}\sum_{i=1}^{T}K'''(\widetilde{e})\frac{\varepsilon_{i-1}-\widehat{\varepsilon_{i-1}}}{g}\\
= &\frac{1}{Tg^4}\sum_{i=1}^{T}K'''(\widetilde{e})\mathbf{x_{i-1}}'\left(\widehat{\mathbf{a}}-\mathbf{a}\right)\\
\leq &\max_{u\in\mathbb{R}}\left|K'''(u)\right|\underbrace{\left(\frac{1}{T}\sum_{i=1}^{T}\frac{1}{\sqrt{Tg^8}}\left|\mathbf{x_{i-1}}'\right|\right)}_{=o_p(1)}\underbrace{\sqrt{T}\left|\widehat{\mathbf{a}}-\mathbf{a}\right|}_{=O_p(1)}\\
= &o_p(1),
\end{aligned}
$$

where $\widetilde{e}$ denotes a suitable value between $\frac{e-\widehat{\varepsilon_{i-1}}}{g}$ and $\frac{e-\varepsilon_{i-1}}{g}$.

To prove (iv) it suffices to show

$$\frac{1}{Tg^3}\sum_{i=1}^{T}\left(K''\left(\frac{e-\widehat{\varepsilon_{i-1}}}{g}\right)-K''\left(\frac{e-\varepsilon_{i-1}}{g}\right)\right)\varepsilon_i^2=o_p(1) \tag{5.7}$$

and

$$\frac{1}{Tg^3}\sum_{i=1}^{T}K''\left(\frac{e-\widehat{\varepsilon_{i-1}}}{g}\right)\left(\widehat{\varepsilon_i}^2-\varepsilon_i^2\right)=o_p(1). \tag{5.8}$$

Firstly, a Taylor expansion for the left hand side of (5.7) yields

$$\begin{aligned}
&\frac{1}{Tg^3}\sum_{i=1}^{T}\left(K''\left(\frac{e-\widehat{\varepsilon_{i-1}}}{g}\right)-K''\left(\frac{e-\varepsilon_{i-1}}{g}\right)\right)\varepsilon_i^2\\
=&\frac{1}{Tg^3}\sum_{i=1}^{T}K'''(\widetilde{e})\frac{\varepsilon_{i-1}-\widehat{\varepsilon_{i-1}}}{g}\varepsilon_i^2\\
\leq&\max_{u\in\mathbb{R}}\left|K'''(u)\right|\frac{1}{Tg^4}\sum_{i=1}^{T}\varepsilon_i^2\left|\mathbf{x_{i-1}}'(\widehat{\mathbf{a}}-\mathbf{a})\right|\\
\leq&\max_{u\in\mathbb{R}}\left|K'''(u)\right|\underbrace{\left(\frac{1}{T}\sum_{i=1}^{T}\frac{\varepsilon_i^2}{\sqrt{Tg^8}}\left|\mathbf{x_{i-1}}'\right|\right)}_{=o_p(1)}\underbrace{\sqrt{T}\left|\widehat{\mathbf{a}}-\mathbf{a}\right|}_{=O_p(1)}\\
=&o_p(1),
\end{aligned}$$

where $\widetilde{e}$ denotes a suitable value between $\frac{e-\widehat{\varepsilon_{i-1}}}{g}$ and $\frac{e-\varepsilon_{i-1}}{g}$. Secondly, the left hand side of (5.8) equals

$$\begin{aligned}
&\frac{1}{Tg^3}\sum_{i=1}^{T} K''\left(\frac{e-\widehat{\varepsilon_{i-1}}}{g}\right)\left(\widehat{\varepsilon_i^2}-\varepsilon_i^2\right)\\
\leq &\max_{u\in\mathbb{R}}\left|K''(u)\right|\frac{1}{Tg^3}\sum_{i=1}^{T}\left|\widehat{\varepsilon_i^2}-\varepsilon_i^2\right|\\
\leq &\max_{u\in\mathbb{R}}\left|K''(u)\right|\left(\frac{1}{Tg^3}\sum_{i=1}^{T}\left(\widehat{\varepsilon_i}-\varepsilon_i\right)^2+\frac{2}{Tg^3}\sum_{i=1}^{T}\left|\varepsilon_i\left(\widehat{\varepsilon_i}-\varepsilon_i\right)\right|\right)\\
= &\max_{u\in\mathbb{R}}\left|K''(u)\right|\left(\frac{1}{Tg^3}\sum_{i=1}^{T}\left(\mathbf{x_i}'(\mathbf{a}-\widehat{\mathbf{a}})\right)^2+\frac{2}{Tg^3}\sum_{i=1}^{T}\left|\varepsilon_i\mathbf{x_i}'(\mathbf{a}-\widehat{\mathbf{a}})\right|\right)\\
\leq &\max_{u\in\mathbb{R}}\left|K''(u)\right|\frac{1}{Tg^3}\underbrace{\left(T\sum_{k=0}^{p}(a_k-\widehat{a}_k)^2\right)}_{=O_p(1)}\left(\frac{1}{T}\sum_{i=1}^{T}\underbrace{\sum_{k=0}^{p}x_{i,k}^2}_{=O_p(1)}\right)\\
&+\max_{u\in\mathbb{R}}\left|K''(u)\right|\underbrace{\left(\frac{2}{\sqrt{T^3g^6}}\sum_{i=1}^{T}\left|\varepsilon_i\mathbf{x_i}'\right|\right)}_{=o_p(1)}\underbrace{\sqrt{T}\left|\mathbf{a}-\widehat{\mathbf{a}}\right|}_{=O_p(1)}\\
= &o_p(1).
\end{aligned}$$

□

Lemma 5.7

Suppose that X_t is generated by the model (5.1) satisfying assumptions (A1)-(A2) and (B1)-(B4). Then for all $e\in\mathbb{R}$

$$\frac{h}{T}\sum_{t=1}^{T}K_h^2(e-\widehat{\varepsilon_{t-1}})\left(\widetilde{\sigma_g^2}(\widehat{\varepsilon_{t-1}})\right)^2=\frac{h}{T}\sum_{t=1}^{T}K_h^2(e-\varepsilon_{t-1})\sigma^4(\varepsilon_{t-1})+o_p(1).$$

Proof. Here it suffices to show

$$\frac{h}{T}\sum_{t=1}^{T}\left(K_h^2(e-\widehat{\varepsilon_{t-1}})-K_h^2(e-\varepsilon_{t-1})\right)\sigma^4(\varepsilon_{t-1})=o_p(1) \tag{5.9}$$

and

$$\frac{h}{T}\sum_{t=1}^{T}K_h^2(e-\widehat{\varepsilon_{t-1}})\left(\left(\widetilde{\sigma_g^2}(\widehat{\varepsilon_{t-1}})\right)^2-\sigma^4(\varepsilon_{t-1})\right)=o_p(1). \tag{5.10}$$

Firstly, a Taylor expansion for the left hand side of (5.9) yields

$$
\begin{aligned}
&\frac{h}{T}\sum_{t=1}^{T}\left(K_h^2(e-\widehat{\varepsilon_{t-1}})-K_h^2(e-\varepsilon_{t-1})\right)\sigma^4(\varepsilon_{t-1})\\
=&\frac{1}{Th}\sum_{t=1}^{T}\left(K^2\left(\frac{e-\widehat{\varepsilon_{t-1}}}{h}\right)-K^2\left(\frac{e-\varepsilon_{t-1}}{h}\right)\right)\sigma^4(\varepsilon_{t-1})\\
=&\frac{1}{Th}\sum_{t=1}^{T}\left(\left(K^2\right)'\left(\frac{e-\varepsilon_{t-1}}{h}\right)\frac{\varepsilon_{t-1}-\widehat{\varepsilon_{t-1}}}{h}\right.\\
&\left.+\frac{1}{2}\left(K^2\right)''(\widetilde{e})\left(\frac{\varepsilon_{t-1}-\widehat{\varepsilon_{t-1}}}{h}\right)^2\right)\sigma^4(\varepsilon_{t-1})\\
=&\frac{1}{Th^2}\sum_{t=1}^{T}\left(K^2\right)'\left(\frac{e-\varepsilon_{t-1}}{h}\right)\left(\varepsilon_{t-1}-\widehat{\varepsilon_{t-1}}\right)\sigma^4(\varepsilon_{t-1})\\
&+\frac{1}{2Th^3}\sum_{t=1}^{T}\left(K^2\right)''(\widetilde{e})\left(\varepsilon_{t-1}-\widehat{\varepsilon_{t-1}}\right)^2\sigma^4(\varepsilon_{t-1})\\
\leq&\left(\frac{1}{Th^2}\sum_{t=1}^{T}\left|\varepsilon_{t-1}-\widehat{\varepsilon_{t-1}}\right|\right)\underbrace{\max_{u\in\mathbb{R}}\left|\left(K^2\right)'(u)\right|}_{=O(1)}\underbrace{\max_{u\in\mathbb{R}}\left|\sigma^4(u)\right|}_{=O(1)}\\
&+\left(\frac{1}{2Th^3}\sum_{t=1}^{T}\left(\varepsilon_{t-1}-\widehat{\varepsilon_{t-1}}\right)^2\right)\underbrace{\max_{u\in\mathbb{R}}\left|\left(K^2\right)''(u)\right|}_{=O(1)}\underbrace{\max_{u\in\mathbb{R}}\left|\sigma^4(u)\right|}_{=O(1)}\\
=&o_p(1),
\end{aligned}
$$

where $\widetilde{e}$ denotes a suitable value between $\frac{e-\widehat{\varepsilon_{t-1}}}{h}$ and $\frac{e-\varepsilon_{t-1}}{h}$. We obtain the last equation from

$$
\begin{aligned}
&\frac{1}{Th^2}\sum_{t=1}^{T}\left|\varepsilon_{t-1}-\widehat{\varepsilon_{t-1}}\right|\\
\leq&\frac{1}{\sqrt{Th^4}}\underbrace{\left(\frac{1}{T}\sum_{t=1}^{T}|\mathbf{x_{t-1}}'|\right)}_{=O_p(1)}\underbrace{\sqrt{T}|\widehat{\mathbf{a}}-\mathbf{a}|}_{=O_p(1)}\\
=&O_p\left(\frac{1}{\sqrt{Th^4}}\right)=o_p(1)
\end{aligned}
$$

and

$$\begin{aligned}
&\frac{1}{Th^3}\sum_{t=1}^{T}\left(\varepsilon_{t-1}-\widehat{\varepsilon_{t-1}}\right)^2 \\
\leq &\frac{1}{Th^3}\underbrace{\left(\frac{1}{T}\sum_{t=1}^{T}\sum_{k=0}^{p}x_{t-1,k}^2\right)}_{=O_p(1)}\underbrace{T\sum_{k=0}^{p}(\widehat{a_k}-a_k)^2}_{=O_p(1)} \\
= &O_p\left(\frac{1}{Th^3}\right)=o_p(1).
\end{aligned}$$

Secondly, a Taylor expansion for the left hand side of (5.10) yields

$$\begin{aligned}
&\frac{h}{T}\sum_{t=1}^{T}K_h^2(e-\widehat{\varepsilon_{t-1}})\left(\left(\widetilde{\sigma_g^2}(\widehat{\varepsilon_{t-1}})\right)^2-\sigma^4(\varepsilon_{t-1})\right) \\
= &\frac{1}{Th}\sum_{t=1}^{T}K^2\left(\frac{e-\widehat{\varepsilon_{t-1}}}{h}\right)\left\{\left(\left(\widetilde{\sigma_g^2}(\widehat{\varepsilon_{t-1}})\right)^2-\sigma^4(\widehat{\varepsilon_{t-1}})\right)\right. \\
&\quad\left.+\left(\sigma^4(\widehat{\varepsilon_{t-1}})-\sigma^4(\varepsilon_{t-1})\right)\right\} \\
= &\frac{1}{Th}\sum_{t=1}^{T}K^2\left(\frac{e-\widehat{\varepsilon_{t-1}}}{h}\right)\left\{\left(\widetilde{\sigma_g^2}(\widehat{\varepsilon_{t-1}})-\sigma^2(\widehat{\varepsilon_{t-1}})\right)^2\right. \\
&\quad\left.+2\left(\sigma^2(\widehat{\varepsilon_{t-1}})\left(\widetilde{\sigma_g^2}(\widehat{\varepsilon_{t-1}})-\sigma^2(\widehat{\varepsilon_{t-1}})\right)\right)\right\} \\
&+\frac{1}{Th}\sum_{t=1}^{T}K^2\left(\frac{e-\widehat{\varepsilon_{t-1}}}{h}\right)\left(\left(\sigma^4\right)'(\varepsilon_{t-1})\left(\widehat{\varepsilon_{t-1}}-\varepsilon_{t-1}\right)\right. \\
&\quad\left.+\frac{1}{2}\left(\sigma^4\right)''(\widetilde{e})\left(\widehat{\varepsilon_{t-1}}-\varepsilon_{t-1}\right)^2\right),
\end{aligned}$$

where $\widetilde{e}$ denotes a suitable value between $\widehat{\varepsilon_{t-1}}$ and ε_{t-1}. The first and the second summand are smaller than

$$\frac{1}{Th^2}\underbrace{\max_{u\in\mathbb{R}} Th\left(\widetilde{\sigma_g^2}(u)-\sigma^2(u)\right)^2}_{=O_p(1)}\underbrace{\max_{u\in\mathbb{R}}\left|K^2(u)\right|}_{=O(1)}$$
$$+\frac{2}{\sqrt{Th^3}}\underbrace{\max_{u\in\mathbb{R}}\sqrt{Th}\left|\widetilde{\sigma_g^2}(u)-\sigma^2(u)\right|}_{=O_p(1)}\underbrace{\max_{u\in\mathbb{R}}\left|K^2(u)\right|}_{=O(1)}\underbrace{\max_{u\in\mathbb{R}}\left|\sigma^2(u)\right|}_{=O(1)}$$
$$=o_p(1)$$

and

$$\underbrace{\left(\frac{1}{Th}\sum_{t=1}^{T}\left|\widehat{\varepsilon_{t-1}}-\varepsilon_{t-1}\right|\right)}_{=O_p\left(\frac{1}{\sqrt{Th^2}}\right)}\underbrace{\max_{u\in\mathbb{R}}\left|K^2(u)\right|}_{=O(1)}\underbrace{\max_{u\in\mathbb{R}}\left|\left(\sigma^4\right)'(u)\right|}_{=O(1)}$$
$$+\underbrace{\left(\frac{1}{2Th}\sum_{t=1}^{T}\left(\widehat{\varepsilon_{t-1}}-\varepsilon_{t-1}\right)^2\right)}_{=O_p\left(\frac{1}{\sqrt{Th}}\right)}\underbrace{\max_{u\in\mathbb{R}}\left|K^2(u)\right|}_{=O(1)}\underbrace{\max_{u\in\mathbb{R}}\left|\left(\sigma^4\right)''(u)\right|}_{=O(1)}$$
$$=o_p(1),$$

respectively, which shows (5.10). □

Lemma 5.8

Suppose that X_t is generated by the model (5.1) satisfying assumptions (A1)-(A2) and (B1)-(B5). Then the residual bootstrap of section 5.2 has the following asymptotic property for all $e\in\mathbb{R}$:

$$\sqrt{\frac{h}{T}}\sum_{t=1}^{T}K_h(e-\widehat{\varepsilon_{t-1}})\sqrt{\kappa-1}\widetilde{\sigma_g^2}(\widehat{\varepsilon_{t-1}})m_t^* \xrightarrow{d} \mathcal{N}\left(0,\tau^2(e)\right) \quad \textit{in probability.}$$

Proof. We observe

$$\phi_t^*(e) := \sqrt{\frac{h}{T}}K_h\left(e-\widehat{\varepsilon_{t-1}}\right)\sqrt{\kappa-1}\widetilde{\sigma_g^2}(\widehat{\varepsilon_{t-1}})m_t^*, \quad t=1,...,T.$$

For each T the sequence $\phi_1^*,...,\phi_T^*$ is independent because $m_1^*,...,m_T^*$ is independent. Here we obtain

$$\mathscr{E}_*\left(\phi_t^*(e)\right)=\sqrt{\frac{h}{T}}K_h(e-\widehat{\varepsilon_{t-1}})\sqrt{\kappa-1}\widetilde{\sigma_g^2}(\widehat{\varepsilon_{t-1}})\mathscr{E}_*(m_t^*)=0$$

and from Lemma 5.7

$$
\begin{aligned}
s_T^2 :&= \sum_{t=1}^{T} \mathscr{E}_*\left(\phi_t^{*2}(e)\right)\\
&= \frac{h}{T}\sum_{t=1}^{T} K_h^2(e-\widehat{\varepsilon_{t-1}})(\kappa-1)\left(\widetilde{\sigma_g^2}(\widehat{\varepsilon_{t-1}})\right)^2 \underbrace{\mathscr{E}_*(m_t^{*2})}_{=1}\\
&= (\kappa-1)\left(\frac{h}{T}\sum_{t=1}^{T} K_h^2(e-\varepsilon_{t-1})\sigma^4(\varepsilon_{t-1}) + o_p(1)\right)\\
&= (\kappa-1)\frac{h}{T}\sum_{t=1}^{T} K_h^2(e-\varepsilon_{t-1})\sigma^4(\varepsilon_{t-1}) + o_p(1)\\
&\xrightarrow{p} \tau^2(e).
\end{aligned}
$$

Then we have the Lyapounov condition for $\delta = 1$

$$
\begin{aligned}
&\sum_{t=1}^{T} \frac{1}{s_T^3}\mathscr{E}_*\left|\phi_t^*(e)\right|^3\\
=&\frac{(\kappa-1)^{3/2}\mathscr{E}_*\left|m_t^*\right|^3}{s_T^3}\sqrt{\frac{h^3}{T}}\left(\frac{1}{T}\sum_{t=1}^{T}\left|K_h(e-\widehat{\varepsilon_{t-1}})\widetilde{\sigma_g^2}(\widehat{\varepsilon_{t-1}})\right|^3\right)\\
=&\frac{(\kappa-1)^{3/2}\mathscr{E}_*\left|m_t^*\right|^3}{s_T^3}\frac{1}{\sqrt{Th^3}}\left(\frac{1}{T}\sum_{t=1}^{T}K^3\left(\frac{e-\widehat{\varepsilon_{t-1}}}{h}\right)\left|\widetilde{\sigma_g^2}(\widehat{\varepsilon_{t-1}})\right|^3\right)\\
=&\frac{(\kappa-1)^{3/2}\mathscr{E}_*\left|m_t^*\right|^3}{s_T^3}\frac{1}{\sqrt{Th^3}}\underbrace{\max_{u\in\mathbb{R}}K^3(u)}_{=O_p(1)}\left(\frac{1}{T}\sum_{t=1}^{T}\underbrace{\left|\widetilde{\sigma_g^2}(\widehat{\varepsilon_{t-1}})\right|^3}_{=O_p(\log^3 T)}\right)\\
=&O_p\left(\frac{\log^3 T}{\sqrt{Th^3}}\right)\\
=&o_p(1).
\end{aligned}
$$

We obtain the last equation from (B5) and

$$
\begin{aligned}
&\widetilde{\sigma_g^2}(\widehat{\varepsilon_{t-1}}) \\
=&\frac{\sum_{i=1}^T K_g(\widehat{\varepsilon_{t-1}}-\varepsilon_{i-1})\widehat{\varepsilon_i^2}}{\sum_{j=1}^T K_g(\widehat{\varepsilon_{t-1}}-\varepsilon_{j-1})} \\
\leq&\max_{1\leq i\leq T}\widehat{\varepsilon_i^2} \\
\leq&2\max_{1\leq i\leq T}\left(\varepsilon_i^2+(\widehat{\varepsilon}_i-\varepsilon_i)^2\right) \\
\leq&2\underbrace{\max_{1\leq i\leq T}\varepsilon_i^2}_{=O_p(\log T)}+2\max_{1\leq i\leq T}\left(\sum_{k=0}^p x_{i,k}(a_k-\widehat{a_k})\right)^2 \\
\leq&O_p(\log T)+2\max_{1\leq i\leq T}\sum_{k=0}^p x_{i,k}^2\sum_{k=0}^p(a_k-\widehat{a_k})^2 \\
\leq&O_p(\log T)+2\underbrace{\left(\frac{1}{T}\sum_{i=1}^T\sum_{k=0}^p x_{i,k}^2\right)}_{=O_p(1)}\underbrace{\left(T\sum_{k=0}^p(a_k-\widehat{a_k})^2\right)}_{=O_p(1)} \\
=&O_p(\log T).
\end{aligned}
$$

The central limit theorem for triangular arrays applied to the sequence $\phi_1^*,\dots,\phi_T^*$ shows

$$\sum_{t=1}^T\phi_t^*(e)\xrightarrow{d}\mathscr{N}\left(0,\tau^2(e)\right)\quad\textit{in probability.}$$

□

Lemma 5.9

Suppose that X_t is generated by the model (5.1) satisfying assumptions (A1)-(A2) and (B1)-(B4). Then for all $e\in\mathbb{R}$

$$\sqrt{\frac{h}{T}}\sum_{t=1}^T K_h(e-\widehat{\varepsilon_{t-1}})\left(\widetilde{\sigma_g^2}(\widehat{\varepsilon_{t-1}})-\widetilde{\sigma_g^2}(e)\right)\xrightarrow{p}b(e).$$

Proof. The left hand side can be decomposed into the following three summands:

$$\sqrt{\frac{h}{T}}\sum_{t=1}^{T} K_h(e-\widehat{\varepsilon_{t-1}})\left(\widetilde{\sigma_g^2}(\widehat{\varepsilon_{t-1}})-\widetilde{\sigma_g^2}(e)\right)$$
$$=\sqrt{\frac{h}{T}}\sum_{t=1}^{T} K_h(e-\varepsilon_{t-1})\left(\widetilde{\sigma_g^2}(\varepsilon_{t-1})-\widetilde{\sigma_g^2}(e)\right)$$
$$+\sqrt{\frac{h}{T}}\sum_{t=1}^{T} K_h(e-\varepsilon_{t-1})\left(\widetilde{\sigma_g^2}(\widehat{\varepsilon_{t-1}})-\widetilde{\sigma_g^2}(\varepsilon_{t-1})\right)$$
$$+\sqrt{\frac{h}{T}}\sum_{t=1}^{T} \left(K_h(e-\widehat{\varepsilon_{t-1}})-K_h(e-\varepsilon_{t-1})\right)\left(\widetilde{\sigma_g^2}(\widehat{\varepsilon_{t-1}})-\widetilde{\sigma_g^2}(e)\right).$$

From Lemma 5.3 and 5.6 we obtain

$$\sqrt{\frac{h}{T}}\sum_{t=1}^{T} K_h(e-\varepsilon_{t-1})\left(\widetilde{\sigma_g^2}(\varepsilon_{t-1})-\widetilde{\sigma_g^2}(e)\right)\xrightarrow{p} b(e).$$

The second summand equals

$$
\begin{aligned}
&\sqrt{\frac{h}{T}}\sum_{t=1}^{T} K_h(e-\varepsilon_{t-1})\Big(\widetilde{\sigma_g^2}(\widehat{\varepsilon_{t-1}})-\widetilde{\sigma_g^2}(\varepsilon_{t-1})\Big)\\
=&\frac{1}{\sqrt{Th}}\sum_{t=1}^{T} K\left(\frac{e-\varepsilon_{t-1}}{h}\right)\bigg(\Big(\widetilde{\sigma_g^2}(\widehat{\varepsilon_{t-1}})-\sigma^2(\widehat{\varepsilon_{t-1}})\Big)\\
&\quad-\Big(\widetilde{\sigma_g^2}(\varepsilon_{t-1})-\sigma^2(\varepsilon_{t-1})\Big)+\Big(\sigma^2(\widehat{\varepsilon_{t-1}})-\sigma^2(\varepsilon_{t-1})\Big)\bigg)\\
=&\frac{1}{T}\sum_{t=1}^{T}\frac{1}{\sqrt{gh}}K\left(\frac{e-\varepsilon_{t-1}}{h}\right)\sqrt{Tg}\Big(\widetilde{\sigma_g^2}(\widehat{\varepsilon_{t-1}})-\sigma^2(\widehat{\varepsilon_{t-1}})\Big)\\
&\quad-\frac{1}{T}\sum_{t=1}^{T}\frac{1}{\sqrt{gh}}K\left(\frac{e-\varepsilon_{t-1}}{h}\right)\sqrt{Tg}\Big(\widetilde{\sigma_g^2}(\varepsilon_{t-1})-\sigma^2(\varepsilon_{t-1})\Big)\\
&\quad+\frac{1}{T}\sum_{t=1}^{T}\frac{1}{\sqrt{gh}}K\left(\frac{e-\varepsilon_{t-1}}{h}\right)\sqrt{Tg}\Big(\sigma^2(\widehat{\varepsilon_{t-1}})-\sigma^2(\varepsilon_{t-1})\Big)\bigg)\\
\leq&\left(\frac{1}{T}\sum_{t=1}^{T}\frac{1}{\sqrt{gh}}K\left(\frac{e-\varepsilon_{t-1}}{h}\right)\right)\underbrace{\max_{u\in\mathbb{R}}\Big(\sqrt{Tg}\Big|\widetilde{\sigma_g^2}(u)-\sigma^2(u)\Big|\Big)}_{=O_p(1)}\\
&\quad+\left(\frac{1}{T}\sum_{t=1}^{T}\frac{1}{\sqrt{gh}}K\left(\frac{e-\varepsilon_{t-1}}{h}\right)\right)\underbrace{\max_{u\in\mathbb{R}}\Big(\sqrt{Tg}\Big|\widetilde{\sigma_g^2}(u)-\sigma^2(u)\Big|\Big)}_{=O_p(1)}\\
&\quad+\frac{1}{T}\sum_{t=1}^{T}\frac{1}{\sqrt{gh}}K\left(\frac{e-\varepsilon_{t-1}}{h}\right)\sqrt{Tg}\Big(\sigma^2\Big)'(\tilde{e})(\widehat{\varepsilon_{t-1}}-\varepsilon_{t-1})\\
\leq&\left(\frac{1}{T}\sum_{t=1}^{T}\frac{1}{\sqrt{gh}}K\left(\frac{e-\varepsilon_{t-1}}{h}\right)\right)\underbrace{\max_{u\in\mathbb{R}}\Big(\sqrt{Tg}\Big|\widetilde{\sigma_g^2}(u)-\sigma^2(u)\Big|\Big)}_{=O_p(1)}\\
&\quad+\left(\frac{1}{T}\sum_{t=1}^{T}\frac{1}{\sqrt{h}}K\left(\frac{e-\varepsilon_{t-1}}{h}\right)|\mathbf{x_{t-1}}'|\right)\underbrace{\sqrt{T}|\mathbf{a}-\hat{\mathbf{a}}|}_{=O_p(1)}\underbrace{\max_{u\in\mathbb{R}}\Big|\Big(\sigma^2\Big)'(u)\Big|}_{=O_p(1)}\\
=&o_p(1).
\end{aligned}
$$

Now let

$$Z_{t-1}:=\frac{1}{\sqrt{gh}}K\left(\frac{e-\varepsilon_{t-1}}{h}\right),$$

then on the one hand

$$
\begin{aligned}
&\frac{1}{T}\mathscr{E}\left(Z_{t-1}^2\right)\\
=&\frac{1}{Tgh}\mathscr{E}\left(K^2\left(\frac{e-\varepsilon_{t-1}}{h}\right)\right)\\
\leq&\frac{1}{Tgh}\max_{u\in\mathbb{R}}K^2(u)\\
=&o(1),
\end{aligned}
$$

thus Remark B.1 yields

$$
\frac{1}{T}\sum_{t=1}^{T}Z_{t-1}=\frac{1}{T}\sum_{t=1}^{T}\mathscr{E}\left(Z_{t-1}\middle|\mathscr{F}_{t-2}\right)+o_p(1).
$$

On the other hand, we obtain

$$
\begin{aligned}
&\frac{1}{T}\sum_{t=1}^{T}\mathscr{E}\left(Z_{t-1}\middle|\mathscr{F}_{t-2}\right)\\
=&\frac{1}{T}\sum_{t=1}^{T}\mathscr{E}\left(\frac{1}{\sqrt{gh}}K\left(\frac{e-\varepsilon_{t-1}}{h}\right)\middle|\mathscr{F}_{t-1}\right)\\
=&\frac{1}{T}\sum_{t=1}^{T}\frac{1}{\sqrt{gh}}\int K\left(\frac{e-u}{h}\right)p(u)du\\
=&-\frac{1}{T}\sum_{t=1}^{T}\sqrt{\frac{h}{g}}\int K(v)p(e-hv)dv\\
=&-\frac{1}{T}\sum_{t=1}^{T}\sqrt{\frac{h}{g}}\int K(v)\left(p(e)-p'(e)hv+\frac{1}{2}p''(\tilde{e})h^2v^2\right)dv\\
=&-\underbrace{\sqrt{\frac{h}{g}}}_{=o(1)}p(e)\underbrace{\int K(v)dv}_{=1}+\sqrt{\frac{h^3}{g}}p'(e)\underbrace{\int vK(v)dv}_{=0}-\frac{1}{2}\sqrt{\frac{h^5}{g}}p''(\tilde{e})\underbrace{\int v^2K(v)dv}_{=O(1)}\\
=&o_p(1),
\end{aligned}
$$

and, therefore, we obtain

$$
\frac{1}{T}\sum_{t=1}^{T}Z_{t-1}=o_p(1).
$$

Analogously, we can prove

$$\frac{1}{T}\sum_{t=1}^{T}\frac{1}{\sqrt{h}}K\left(\frac{e-\varepsilon_{t-1}}{h}\right)\left|\mathbf{x_{t-1}}'\right| = o_p(1),$$

which shows that the second summand is $o_p(1)$.

The last summand equals

$$\begin{aligned}
&\sqrt{\frac{h}{T}}\sum_{t=1}^{T}\left(K_h(e-\widehat{\varepsilon_{t-1}})-K_h(e-\varepsilon_{t-1})\right)\left(\widetilde{\sigma_g^2}(\widehat{\varepsilon_{t-1}})-\widetilde{\sigma_g^2}(e)\right)\\
=&\left(\sqrt{\frac{h}{T}}\sum_{t=1}^{T}\left(K_h(e-\widehat{\varepsilon_{t-1}})-K_h(e-\varepsilon_{t-1})\right)\right)\left(\left(\widetilde{\sigma_g^2}(\widehat{\varepsilon_{t-1}})-\sigma^2(\widehat{\varepsilon_{t-1}})\right)\right.\\
&\left.-\left(\widetilde{\sigma_g^2}(e)-\sigma^2(e)\right)+\left(\sigma^2(\widehat{\varepsilon_{t-1}})-\sigma^2(e)\right)\right)\\
\leq&\underbrace{\left(\sqrt{\frac{h}{T}}\sum_{t=1}^{T}\left|K_h(e-\widehat{\varepsilon_{t-1}})-K_h(e-\varepsilon_{t-1})\right|\right)}_{\overset{Lemma\ 5.5}{=}o_p(1)}\\
&\left(\underbrace{\max_{u\in\mathbb{R}}\left(\sqrt{Tg}\left|\widetilde{\sigma_g^2}(u)-\sigma^2(u)\right|\right)}_{=O_p(1)}+\underbrace{\max_{u\in\mathbb{R}}\sigma^2(u)}_{=O_p(1)}\right)\\
=&o_p(1).
\end{aligned}$$

□

Theorem 5.3

Suppose that X_t is generated by the model (5.1) satisfying assumptions (A1)-(A2) and (B1)-(B5). Then the residual bootstrap of section 5.2 is weakly consistent, i.e.

$$\sqrt{Th}\left(\widetilde{\sigma_h^{*2}}(e)-\widetilde{\sigma_g^2}(e)\right)\xrightarrow{d}\mathcal{N}\left(\frac{b(e)}{p(e)},\frac{\tau^2(e)}{p^2(e)}\right)\quad \textit{in probability.}$$

Proof. Lemma 5.5, 5.8 and 5.9 together with the Slutsky theorem yield the desired result. □

5.3 Wild Bootstrap

Analogously to the previous section, we propose a wild bootstrap technique and prove its weak consistency. A wild bootstrap method can be applied to the model (5.1) as follows:

(step 1) Obtain OLS estimator $\widehat{\mathbf{a}}$ and calculate the residuals

$$\widehat{\varepsilon_t} = X_t - \widehat{a_0} - \sum_{i=1}^{p} \widehat{a_i} X_{t-i}, \quad t = 0, ..., T.$$

(step 2) Obtain the NW estimator with a bandwidth g

$$\widetilde{\sigma_g^2}(e) = \frac{T^{-1}\sum_{i=1}^{T} K_g(e - \widehat{\varepsilon_{i-1}})\widehat{\varepsilon_i}^2}{T^{-1}\sum_{j=1}^{T} K_g(e - \widehat{\varepsilon_{j-1}})}.$$

(step 3) Generate the bootstrap process $\varepsilon_t^{\dagger 2}$ by computing

$$\varepsilon_t^{\dagger 2} = \widetilde{\sigma_g^2}(\widehat{\varepsilon_{t-1}}) + \sqrt{\kappa - 1}\widetilde{\sigma_g^2}(\widehat{\varepsilon_{t-1}})w_t^\dagger,$$
$$w_t^\dagger \overset{iid}{\sim} \mathscr{N}(0,1), \quad t = 1, ..., T.$$

(step 4) Build the bootstrap model

$$\varepsilon_t^{\dagger 2} = \sigma^{\dagger 2}(\widehat{\varepsilon_{t-1}}) + \sqrt{\kappa - 1}\sigma^{\dagger 2}(\widehat{\varepsilon_{t-1}})w_t^\dagger$$

and calculate the NW estimator $\widetilde{\sigma_h^{\dagger 2}}(e)$ with another bandwidth h

$$\widetilde{\sigma_h^{\dagger 2}}(e) = \frac{T^{-1}\sum_{i=1}^{T} K_h(e - \widehat{\varepsilon_{i-1}})\varepsilon_i^{\dagger 2}}{T^{-1}\sum_{j=1}^{T} K_h(e - \widehat{\varepsilon_{j-1}})}.$$

Analogously to the previous section, we observe the decomposition

$$\begin{aligned}
&\sqrt{Th}\left(\widetilde{\sigma_h^{\dagger 2}}(e) - \widetilde{\sigma_g^2}(e)\right) \\
&= \frac{\sqrt{\frac{h}{T}}\sum_{i=1}^{T} K_h(e - \widehat{\varepsilon_{i-1}})\sqrt{\kappa - 1}\widetilde{\sigma_g^2}(\widehat{\varepsilon_{i-1}})w_i^\dagger}{\frac{1}{T}\sum_{j=1}^{T} K_h(e - \widehat{\varepsilon_{j-1}})} \\
&\quad + \frac{\sqrt{\frac{h}{T}}\sum_{i=1}^{T} K_h(e - \widehat{\varepsilon_{i-1}})\left(\widetilde{\sigma_g^2}(\widehat{\varepsilon_{i-1}}) - \widetilde{\sigma_g^2}(e)\right)}{\frac{1}{T}\sum_{j=1}^{T} K_h(e - \widehat{\varepsilon_{j-1}})}
\end{aligned}$$

and analyse asymptotic properties of each part separately.

Lemma 5.10

Suppose that X_t is generated by the model (5.1) satisfying assumptions (A1)-(A2) and (B1)-(B4). Then the wild bootstrap of section 5.3 has the following asymptotic property for all $e \in \mathbb{R}$:

$$\sqrt{\frac{h}{T}}\sum_{t=1}^{T} K_h(e-\widehat{\varepsilon_{t-1}})\sqrt{\kappa-1}\widetilde{\sigma_g^2}(\widehat{\varepsilon_{t-1}})w_t^\dagger \xrightarrow{d} \mathcal{N}\left(0,\tau^2(e)\right) \quad \textit{in probability.}$$

Proof. The proof is the mirror image of Lemma 5.8. We observe

$$\phi_t^\dagger(e) := \sqrt{\frac{h}{T}} K_h\left(e-\widehat{\varepsilon_{t-1}}\right)\sqrt{\kappa-1}\widetilde{\sigma_g^2}(\widehat{\varepsilon_{t-1}})w_t^\dagger, \quad t=1,...,T.$$

For each T the sequence $\phi_1^\dagger,...,\phi_T^\dagger$ is independent because $w_1^\dagger,...,w_T^\dagger$ is independent. Here we obtain

$$\mathscr{E}_\dagger\left(\phi_t^\dagger(e)\right) = 0$$

and

$$\begin{aligned} s_T^2 :&= \sum_{t=1}^{T} \mathscr{E}_\dagger\left(\phi_t^{\dagger 2}(e)\right) \\ &= \frac{h}{T}\sum_{t=1}^{T} K_h^2(e-\widehat{\varepsilon_{t-1}})(\kappa-1)\left(\widetilde{\sigma_g^2}(\widehat{\varepsilon_{t-1}})\right)^2 \underbrace{\mathscr{E}_\dagger(w_t^{\dagger 2})}_{=1} \\ &\xrightarrow{p} \tau^2(e). \end{aligned}$$

Then we have the Lyapounov condition for $\delta = 1$

$$\begin{aligned} &\sum_{t=1}^{T} \frac{1}{s_T^3}\mathscr{E}_\dagger\left|\phi_t^\dagger(e)\right|^3 \\ =&\frac{(\kappa-1)^{3/2}\mathscr{E}_\dagger|w_t^\dagger|^3}{s_T^3}\sqrt{\frac{h^3}{T}}\left(\frac{1}{T}\sum_{t=1}^{T}\left|K_h(e-\widehat{\varepsilon_{t-1}})\widetilde{\sigma_g^2}(\widehat{\varepsilon_{t-1}})\right|^3\right) \\ =&o_p(1). \end{aligned}$$

The central limit theorem for triangular arrays applied to the sequence $\phi_1^\dagger,...,\phi_T^\dagger$ shows

$$\sum_{t=1}^{T}\phi_t^\dagger(e) \xrightarrow{d} \mathcal{N}\left(0,\tau^2(e)\right) \quad \textit{in probability.}$$

□

Theorem 5.4

Suppose that X_t is generated by the model (5.1) satisfying assumptions (A1)-(A2) and (B1)-(B4). Then the wild bootstrap of section 5.3 is weakly consistent, i.e.

$$\sqrt{Th}\left(\widetilde{\sigma_h^{\dagger 2}}(e) - \widetilde{\sigma_g^2}(e)\right) \xrightarrow{d} \mathcal{N}\left(\frac{b(e)}{p(e)}, \frac{\tau^2(e)}{p^2(e)}\right) \quad \text{in probability.}$$

Proof. Lemma 5.5, 5.9 and 5.10 together with the Slutsky theorem yield the desired result. □

5.4 Simulations

In this section we demonstrate the large sample properties of the residual bootstrap of section 5.2 and the wild bootstrap of section 5.3. Let us consider the following semiparametric AR(1)-ARCH(1) model:

$$\begin{aligned} X_t &= a_0 + a_1 X_{t-1} + \varepsilon_t, \\ \varepsilon_t &= \sigma(\varepsilon_{t-1})\eta_t, \quad t = 1, ..., 10{,}000, \end{aligned}$$

where $X_0 = \mathscr{E}(X_t) = \frac{a_0}{1-a_1}$.

The simulation procedure is as follows. Note that in (step 5) it would be better to choose some data sets and adopt the bootstrap techniques based on each data set 2,000 times respectively. It is, however, not easy to show the result in a figure. Therefore, we present here one characteristic result for each bootstrap method.

(step 1) Simulate the bias $\eta_t \overset{iid}{\sim} \mathcal{N}(0,1)$ for $t = 1, ..., 10{,}000$.

(step 2) Let $\sigma(\varepsilon_{t-1}) = \sqrt{b_0 + b_1 \varepsilon_{t-1}^2}$, $\varepsilon_0^2 = \mathscr{E}(\varepsilon_t^2) = \frac{b_0}{1-b_1}$, $a_0 = 0.141$, $a_1 = 0.433$, $b_0 = 0.607$, $b_1 = 0.135$ and calculate

$$X_t = a_0 + a_1 X_{t-1} + \left(b_0 + b_1 \varepsilon_{t-1}^2\right)^{1/2} \eta_t$$

for $t = 1, ..., 10{,}000$.

(step 3) From the data set $\{X_1, X_2, ..., X_{10{,}000}\}$ obtain the OLS estimators $\widehat{a_0}$, $\widehat{a_1}$ and then the NW estimators $\widetilde{\sigma_g^2}(e)$, $\widetilde{\sigma_h^2}(e)$, where $e = 0$ with $g = T^{-1/15}$ and $h = T^{-1/5}$ with $T = 10{,}000$.

(step 4) Repeat (step 1)-(step 3) 2,000 times.

(step 5) Choose one data set $\{X_1, X_2, ..., X_{10,000}\}$ and its estimators $\left\{\widetilde{\sigma_g^2}(e), \widetilde{\sigma_h^2}(e)\right\}$. Adopt the residual bootstrap of section 5.2 and the wild bootstrap of section 5.3 based on the chosen data set and estimators. Both bootstrap methods are repeated 2,000 times.

(step 6) Compare the simulated density of the NW estimators ($\sqrt{Th}\left(\widetilde{\sigma_h^2}(e) - \sigma^2(e)\right)$, thin lines) with the residual and the wild bootstrap approximations ($\sqrt{Th}\left(\widetilde{\sigma_h^{*2}}(e) - \widetilde{\sigma_g^2}(e)\right)$ and $\sqrt{Th}\left(\sigma_h^{\dagger 2}(e) - \widetilde{\sigma_g^2}(e)\right)$, bold lines).

Figure 5.1 and Figure 5.2 indicate that the residual and the wild bootstrap are weakly consistent, which corresponds to the theoretical results of Theorem 5.3 and 5.4.

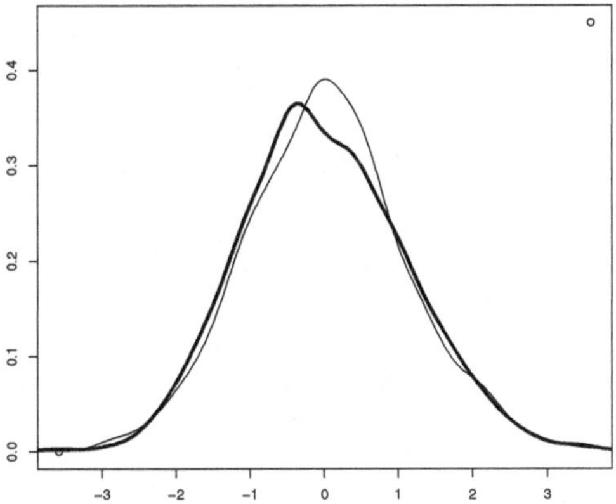

Figure 5.1: Residual Bootstrap

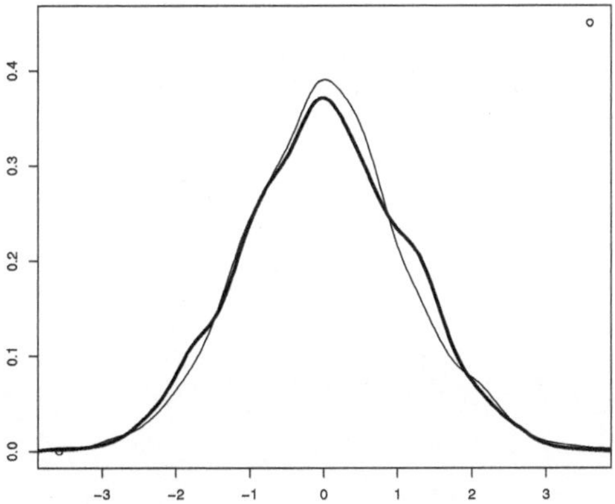

Figure 5.2: Wild Bootstrap

Appendix

A Central Limit Theorems

Definition A.1 (Martingale Difference Sequence)
Let $\{Z_k, k \in \mathbb{N}\}$ be a sequence of integrable random variables defined on a common probability space $(\Omega, \mathscr{A}, P)$ and let $\{\mathscr{F}_k, k \in \mathbb{N} \bigcup \{0\}\}$ denote an increasing sequence of sub-σ-fields of $\mathscr{A}$ such that Z_k is $\mathscr{F}_k$-measurable for each $k \in \mathbb{N}$. Then $\{Z_k\}$ is called a martingale difference sequence if $\mathscr{E}(Z_k|\mathscr{F}_{k-1}) = 0$ almost surely for every $k \in \mathbb{N}$.

Theorem A.1
Let $\{Z_k\}$ be a martingale difference sequence with

$$0 < s_K^2 := \sum_{k=1}^{K} \mathscr{E}(Z_k^2) < \infty.$$

Then

$$s_K^{-1} \sum_{k=1}^{K} Z_k \xrightarrow{d} \mathscr{N}(0,1)$$

if

$$(a) \quad s_K^{-2} \sum_{k=1}^{K} \mathscr{E}(Z_k^2|\mathscr{F}_{k-1}) \xrightarrow{P} 1 \quad and$$

$$(b) \quad s_K^{-2} \sum_{k=1}^{K} \mathscr{E}\left(Z_k^2 \mathbf{1}\{|Z_k| \geq \varepsilon s_K\}\right) \xrightarrow{P} 0 \quad for\ every\ \varepsilon > 0.$$

Proof. Brown (1971), Theorem 1. □

Here we observe a triangular array $\{Z_{n,k}, 1 \leq k \leq k_n, n \in \mathbb{N}\}$ of integrable random variables defined on a common probability space $(\Omega, \mathscr{A}, P)$. Let $\{\mathscr{F}_{n,k}, 1 \leq k \leq k_n, n \in \mathbb{N}\}$ be a given array of sub-σ-fields of $\mathscr{A}$ such that $Z_{n,k}$ is $\mathscr{F}_{n,k}$-measurable and $\mathscr{F}_{n,k}$ is monotone increasing in k for every n.

Theorem A.2
Suppose that for each n the sequence $Z_{n,1}, \ldots, Z_{n,k_n}$ is independent, $\mathscr{E}(Z_{n,k}) = 0$ and $0 < s_n^2 := \sum_{k=1}^{k_n} \mathscr{E}(Z_{n,k}^2) < \infty$. If the Lindeberg condition

$$\lim_{n \to \infty} \sum_{k=1}^{k_n} \frac{1}{s_n^2} \mathscr{E}\left(Z_{n,k}^2 \mathbf{1}\left\{|Z_{n,k}| \geq \varepsilon s_n\right\}\right) = 0$$

holds for all positive ε, then

$$s_n^{-1} \sum_{k=1}^{k_n} Z_{n,k} \xrightarrow[n\to\infty]{d} \mathcal{N}(0,1).$$

Proof. Billingsley (1995), Theorem 27.2. □

Corollary A.1
If, under the assumptions of Theorem A.2, the Lyapounov condition

$$\lim_{n\to\infty} \sum_{k=1}^{k_n} \frac{1}{s_n^{2+\delta}} \mathscr{E}\left(|Z_{n,k}|^{2+\delta}\right) = 0$$

holds for some positive δ, then the Lindeberg condition holds for all positive ε.

Proof. Billingsley (1995), Theorem 27.3. □

B Miscellanea

Proposition B.1

Given $\mathbf{g} : \mathbb{R}^k \to \mathbb{R}^l$ and any sequence $\{\mathbf{b}_n\}$ of $k \times 1$ random vectors such that $\mathbf{b}_n \xrightarrow{p} \mathbf{b}$, where $\mathbf{b}$ is a $k \times 1$ vector. If $\mathbf{g}$ is continuous at $\mathbf{b}$, then $\mathbf{g}(\mathbf{b}_n) \xrightarrow{p} \mathbf{g}(\mathbf{b})$.

Proof. White (2001), Proposition 2.27. □

Remark B.1

For every sequence $\{Z_t,\ t = 1,...,T\}$ of random variables with

$$\frac{\mathscr{E}(Z_t^2)}{T} = o(1)$$

we obtain

$$\begin{aligned}
&\mathscr{E}\left[\frac{1}{T}\sum_{t=1}^{T}\left\{Z_t - \mathscr{E}\left(Z_t \middle| \mathscr{F}_{t-1}\right)\right\}\right]^2 \\
=&\frac{1}{T^2}\sum_{s,\,t=1}^{T}\mathscr{E}\left[\left\{Z_t - \mathscr{E}\left(Z_t \middle| \mathscr{F}_{t-1}\right)\right\}\left\{Z_s - \mathscr{E}\left(Z_s \middle| \mathscr{F}_{s-1}\right)\right\}\right] \\
=&\frac{1}{T^2}\sum_{t=1}^{T}\underbrace{\mathscr{E}\left\{Z_t - \mathscr{E}\left(Z_t \middle| \mathscr{F}_{t-1}\right)\right\}^2}_{\leq \mathscr{E}(Z_t^2)} \\
=&o(1).
\end{aligned}$$

Chebychev's inequality yields for every $\delta > 0$

$$\begin{aligned}
&P\left\{\left|\frac{1}{T}\sum_{t=1}^{T} Z_t - \frac{1}{T}\sum_{t=1}^{T}\mathscr{E}\left(Z_t \middle| \mathscr{F}_{t-1}\right)\right| > \delta\right\} \\
\leq&\frac{1}{\delta^2}\mathscr{E}\left[\frac{1}{T}\sum_{t=1}^{T}\left\{Z_t - \mathscr{E}\left(Z_t \middle| \mathscr{F}_{t-1}\right)\right\}\right]^2 \\
=&o(1),
\end{aligned}$$

that is, according to the definition

$$\frac{1}{T}\sum_{t=1}^{T} Z_t \xrightarrow{p} \frac{1}{T}\sum_{t=1}^{T}\mathscr{E}\left(Z_t \middle| \mathscr{F}_{t-1}\right).$$

Theorem B.1
Let $\{Z_t, \mathscr{F}_t\}$ be a martingale difference sequence. If

$$\sum_{t=1}^{\infty} \frac{\mathscr{E}(|Z_t|^{2+p})}{t^{1+p}} < \infty$$

for some $p \geq 1$, then

$$\frac{1}{T} \sum_{t=1}^{T} Z_t \xrightarrow{a.s.} 0.$$

Proof. Stout (1974, p. 154-155). □

Bibliography

[1] Basel Committee on Banking Supervision (1996) *Amendment to the Capital Accord to incorporate market risks*. Bank for International Settlements, Basel.

[2] Berkes, I., Horváth, L. and Kokoszka, P. (2003) GARCH processes: structure and estimation. *Bernoulli* **9**, 201-227.

[3] Billingsley, P. (1995) *Probability and Measure,* 3rd Edition. Wiley, New York.

[4] Bollerslev, T. (1986) Generalized autoregressive conditional heteroskedasticity. *Journal of Econometrics* **31**, 307-327.

[5] Bollerslev, T., Chou, R. Y. and Kroner, K. F. (1992) ARCH modeling in finance. *Journal of Econometrics* **52**, 5-59.

[6] Brown, B. M. (1971) Martingale Central Limit Theorems. *The Annals of Mathematical Statistics* **42**, 59-66.

[7] Chang, B. R. (2006) Applying nonlinear generalized autoregressive conditional heteroscedasticity to compensate ANFIS outputs tuned by adaptive support vector regression. *Fuzzy Sets and Systems* **157**, 1832-1850.

[8] Comte, F. and Lieberman, O. (2003) Asymptotic theory for multivariate GARCH processes. *Journal of Multivariate Analysis* **84**, 61-84.

[9] Dowd, K. (2005) *Measuring market risk,* 2nd Edition. Wiley, Chichester.

[10] Efron, B. (1979) Bootstrap methods: another look at the jackknife. *The Annals of Statistics* **7**, 1-26.

[11] Efron, B. and Tibshirani, R. J. (1993) *An Introduction to the Bootstrap*. Chapman & Hall, New York.

[12] Engle, R. F. (1982) Autoregressive conditional heteroscedasticity with estimates of the variance of United Kingdom inflation. *Econometrica* **50**, 987-1007.

[13] Engle, R. F. and González-Rivera, G. (1991) Semiparametric ARCH Models. *Journal of Business & Economic Statistics* **9**, 345-359.

[14] Fama, E. F. (1965) The Behavior of Stock-Market Prices. *The Journal of Business* **38**, 34-105.

[15] Fan, J. and Yao, Q. (1998) Efficient estimation of conditional variance functions in stochastic regression. *Biometrika* **85**, 645-660.

[16] Francq, C. and Zakoïan, J.-M. (2004) Maximum likelihood estimation of pure GARCH and ARMA-GARCH processes. *Bernoulli* **10**, 605-637.

[17] Franke, J., Kreiß, J.-P. and Mammen, E. (1997) *Bootstrap of kernel smoothing in nonlinear time series.* Working Paper, Technische Universität Braunschweig.

[18] Franke, J., Kreiß, J.-P. and Mammen, E. (2002) Bootstrap of kernel smoothing in nonlinear time series. *Bernoulli* **8**, 1-37.

[19] Fuller, W. A. (1996) *Introduction to Statistical Time Series,* 2nd Edition. Wiley, New York.

[20] Hartz, C., Mittnik, S. and Paolella, M. (2006) Accurate value-at-risk forecasting based on the normal-GARCH model. *Computational Statistics & Data Analysis* **51**, 2295-2312.

[21] Horowitz, J. L. (2003) The Bootstrap in Econometrics. *Statistical Science* **18**, 211-218.

[22] Jorion, P. (2007) *Value at Risk: The New Benchmark for Managing Financial Risk,* 3rd Edition. McGraw-Hill, New York.

[23] Kreiß, J.-P. (1997) *Asymptotical Properties of Residual Bootstrap for Autoregressions.* Working Paper, Technische Universität Braunschweig.

[24] Kreiß, J.-P. and Franke, J. (1992) Bootstrapping stationary autoregressive moving-average models. *Journal of Time Series Analysis* **13**, 297-317.

[25] Lee, S.-W. and Hansen, B. E. (1994) Asymptotic theory for the GARCH(1,1) quasi-maximum likelihood estimator. *Econometric Theory* **10**, 29-52.

[26] Ling, S. (2007) Self-weighted and local quasi-maximum likelihood estimators for ARMA-GARCH/IGARCH models. *Journal of Econometrics* **140**, 849-873.

[27] Ling, S. and Li, W. K. (1997) On Fractionally Integrated Autoregressive Moving-Average Time Series Models With Conditional Heteroscedasticity. *Journal of the American Statistical Association* **92**, 1184-1194.

[28] Ling, S. and McAleer, M. (2003a) Asymptotic theory for a vector ARMA-GARCH model. *Econometric Theory* **19**, 280-310.

[29] Ling, S. and McAleer, M. (2003b) On Adaptive Estimation in Nonstationary ARMA Models with GARCH Errors. *The Annals of Statistics* **31**, 642-674.

[30] Liu, R. Y. (1988) Bootstrap procedures under some non-i.i.d. models. *The Annals of Statistics* **16**, 1696-1708.

[31] Lumsdaine, R. L. (1996) Consistency and Asymptotic Normality of the Quasi-Maximum Likelihood Estimator in IGARCH(1, 1) and Covariance Stationary GARCH(1, 1) Models. *Econometrica* **64**, 575-596.

[32] Maercker, G. (1997) *Statistical Inference in Conditional Heteroskedastic Autoregressive Models.* PhD Dissertation, Technische Universität Braunschweig.

[33] Mammen, E. (1992) *When Does Bootstrap Work?: Asymptotic Results and Simulations,* Lecture Notes in Statistics 77. Springer, New York.

[34] Mandelbrot, B. (1963) The Variation of Certain Speculative Prices. *The Journal of Business* **36**, 394-419.

[35] Nicholls, D. F. and Pagan, A. R. (1983) Heteroscedasticity in Models with Lagged Dependent Variables. *Econometrica* **51**, 1233-1242.

[36] Palkowski, F. (2005) *Nichtparametrisches Bootstrap in heteroskedastischen Zeitreihen*. PhD Dissertation, Technische Universität Braunschweig.

[37] Pantula, S. G. (1988) Estimation of autoregressive models with ARCH errors. *Sankhyā: The Indian Journal of Statistics B* **50**, 119-138.

[38] Pascual, L., Romo, J. and Ruiz, E. (2005) Bootstrap prediction for returns and volatilities in GARCH models. *Computational Statistics & Data Analysis* **50**, 2293-2312.

[39] Reeves, J. J. (2005) Bootstrap prediction intervals for ARCH models. *International Journal of Forecasting* **21**, 237-248.

[40] Robio, P. O. (1999) Forecast intervals in ARCH models: bootstrap versus parametric methods. *Applied Economics Letters* **6**, 323-327.

[41] Shao, J. and Tu, D. (1996) *The jackknife and bootstrap*. Springer, Berlin.

[42] Stout, W. F. (1974) *Almost Sure Convergence.* Academic Press, New York.

[43] Straumann, D. (2005) *Estimation in Conditionally Heteroscedastic Time Series Models,* Lecture Notes in Statistics 181. Springer, Berlin.

[44] Tapia, R. A. and Thompson, J. R. (1978) *Nonparametric probability density estimation*. Johns Hopkins University Press, Baltimore.

[45] Thombs, L. A. and Schucany, W. R. (1990) Bootstrap Prediction Intervals for Autoregression. *Journal of the American Statistical Association* **85**, 486-492.

[46] Weiss, A. A. (1984) ARMA models with ARCH errors. *Journal of Time Series Analysis* **5**, 129-143.

[47] Weiss, A. A. (1986) Asymptotic theory for ARCH models: Estimation and testing. *Econometric Theory* **2**, 107-131.

[48] White, H. (2001) *Asymptotic Theory for Econometricians,* Revised Edition. Academic Press, San Diego.

Index

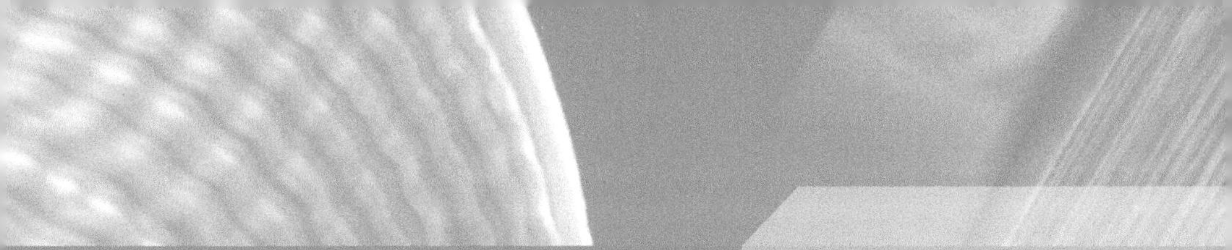